THE TECHNOLOGICAL IMAGINATION

Theories of Contemporary Culture, Volume 3
Center for Twentieth Century Studies
General Editor, Thomas Ewens

THE TECHNOLOGICAL IMAGINATION: Theories and Fictions

edited by

TERESA DE LAURETIS

ANDREAS HUYSSEN

KATHLEEN WOODWARD

Coda Press, Inc.

Madison, Wisconsin

Printed in the United States of America

A co-publication of The Center for Twentieth Century Studies, University of Wisconsin, Milwaukee, Wisconsin & Coda Press, Inc., 700 West Badger Road, Madison, WI 53713

ISBN 0-930956-11-7 (pbk.)

Library of Congress Cataloging in Publication Data
Main entry under title:

The Technological imagination.

(Theories of Contemporary Culture ; 3)
Bibliography: p. 195
1. Technology and civilization—Addresses, essays, lectures. 2. Technology—Social aspects—Addresses, essays, lectures. I. De Lauretis, Teresa. II. Huyssen, Andreas. III. Woodward, Kathleen M. IV. Series.
HM221.T398 303.4'83 80-18718

TABLE OF CONTENTS

Foreword

All too often technology and the imagination have been considered mutually exclusive, or at best incommensurable spheres of social endeavor. Technology is usually understood to be part of the domain of economic and functional efficiency, related to rationality, science, and industrial production, while the imagination is attributed to the humanistic, or humane, aspect of culture—the fine arts, literature, and entertainment. This dichotomy of the "two cultures," as C. P. Snow called it, has powerful roots in the history of Western civilization and in the structure of capitalist society, and cannot be as easily overcome as the frequently voiced demands for a convergence seem to suggest.

Since the beginnings of the industrial revolution in the eighteenth century, the difference between technology and science on the one hand and art, literature, and the imagination on the other has widened into a seemingly unbridgeable chasm. To be sure, this reflects increasing specialization of capitalist societies. But while it can be argued that specialization was necessary and unavoidable, it did give rise to the ideologically loaded concepts of the autonomy of art and the autonomy of technology. To some extent, these concepts have meant different things in different places at different times. Nevertheless, one of their major effects since the nineteenth century has been to legitimize the mechanisms of social and political control. By insisting on the autonomy of art and literature, society has relegated the artistic imagination to the margins and virtually censored attempts by artists to leave the ghetto of high culture and intervene in social and political matters. As a result, the critical impulse of art and literature has been neutralized, and the principles of efficiency and instrumental rationality have taken hold. The concept of autonomous technology has served a similar purpose, but in a different way. Autonomous technology conjures up the nightmare of technics spinning out of control, the horrible vision of Prometheus turning into Frankenstein. At the same time, however, the idea that technology follows its own inherent laws of development has become an ideological and political force, a means of supporting new technologies at a time when our technological civilization is increasingly under attack.

This book challenges the artificial separation of technology from the imagination and argues that the pervasive technologization of everyday life, since the beginning of this century at least, has shaped and transformed all cultural processes from the ways in which we communicate with each other to the ways in which we perceive ourselves and the world. Values, beliefs, and attitudes—both personal and social—are inextricably bound up with technical developments. For as we propose, modern technology cannot be thought of as merely a set of tools or techniques which are "used" to produce objects or commodities. To an ever increasing degree, industrial and communication technologies are determining the nature of work, and leisure-time technologies, as they become available to more and more people, are affecting the private sphere. Therefore, we must begin to understand technology as a relation of the technical and the cultural, as a material and cognitive form of social process.

Technology, or better, *technologies* are intrinsic components of cultural transformation. Thus, literary and artistic production, even when not strictly dependent on the development of special technical means, reflects that cultural transformation and the technologies that effect it. The volume's title—*The Technological Imagination: Theories and Fictions*—does not imply the existence of a *non*-technical imagination. On the contrary, it suggests that technology shapes the very content and form of the imagination in our time and, further, that the notion of imagination cannot be detached from the discourses and practices, the theories and fictions, in which it is concretely, textually inscribed.

This book, then, is conceived as an exploration of the relationship between technology, literature, and culture according to the assumptions outlined above. While much work along these lines has been done in other areas, such as cinema, where the technological nature of the medium has required theorists and critics to address this question, such research has not been undertaken in any major way in literary studies. The three sections that comprise the book are not intended to provide exhaustive surveys of either the areas or the critical approaches in question. Rather they confront some of the problems in understanding how technology is reflected in literature, or what might be called the literary imagination.

The first section, *Technology and American Culture,* analyzes the conflict between authoritarian and democratic technologies since Jefferson, the impact of photography on the aesthetics of realism and the literary responses to and critical assessments of twentieth-century technology. *Machines, Myths, and Marxism* deals with the conceptions of technology in Marxian theory, in the Soviet avant-garde of the early 1920s, and in authors ranging from Ernst Jünger and Bertolt Brecht to contemporary East German writers. Together the essays voice a cutting critique of orthodox Marxism's reliance on technology as a tool for political emancipation. *Science Fictions* considers science fiction historically as a literary genre which uses scientific models of cognition to construct fictional narratives set in technological environments. And, because science fiction is an aesthetic form, a verbal art, it must also be assessed in terms of the practices of writing and reading that it promotes. The essays in this section advance a critical discourse that seeks to account for the textual inscription of technology in cultural processes.

The contributors to this volume take an historical approach and reject the standard view of technology as merely tools or machinery. They also question literary criticism's exclusive concern with the text and nothing but the text. In addition, all of the essays either explicitly or implicitly refuse the traditional *Kulturkritik* that pretends to preserve literature from the threats of urbanization and technologization. For it is by abandoning the view that literature is or should be autonomous, that is, unrelated to everyday life in either its production or its reception, that we can begin to study how our technological environment has affected every aspect of literary discourse—not only content but also form—and the very practice of writing and reading itself.

Understanding the impact on literature of different technologies, however, does not entail an uncritical acceptance or glorification of technology. On the contrary, the authors share a basic skepticism of the ideology of technological progress. At a time when the dominant principles underlying modern civilization—industrial and economic growth, progress as continuing technological development, instrumental rationality, and the performance principle—are fast becoming problematic if not obsolete, it is essential for humanists to explore the ways in which various technologies have been perceived, how they have entered into fictional and theoretical discourses, and how they have shaped literary representations of everyday life.

The critique of technology as autonomous, as progress, that is present in many of these essays does not represent a return to the monolithic condemnations of technology found, for example, in Ellul's *The Technological Society*, Mumford's *The Pentagon of Power*, or Marcuse's *One-Dimensional Man*. Instead the authors try to come to terms with different technologies as perceived by writers in different societies and periods in recent history. This new critique of technology is also expressed in the materials examined. It is exemplified in the changing attitudes of American writers and critics toward technology discussed in the first section. It is evident in the move toward a critique of technology in the literature of Eastern European countries, where the sanctity of technology has been taken for granted as much, if not more, than in the West. And it appears quite clearly in the break of recent American science fiction with the narrative strategies of simple technological extrapolation and the previous emphasis on the hard sciences.

The essays assembled in this volume result from research conducted in conjunction with the Center for Twentieth Century Studies at the University of Wisconsin-Milwaukee in 1977–78. The interdisciplinary study of the relationship between culture and technology in industrial societies has been a major focus of the Center's programs and research for several years. The framework for this research, including the work leading to this book, was initiated and guided by the Center's late Director, Michel Benamou, to whom we are deeply indebted.

We are grateful to the Center for Twentieth Century Studies and its staff, especially Mary Lydon, Publications Coordinator, and Carol Tennessen, Program Coordinator. We would also like to thank Penny Green, Jean Lile, and Trudy Sensat for typing and copy-editing the manuscript. We owe special thanks to Charles Caramello, who distilled and edited the materials left by Michel Benamou. Finally, we acknowledge the College of Letters and Science and the Graduate School of the University of Wisconsin-Milwaukee for contributing their resources and support to this project.

The Editors

I
Technology and American Culture

Introduction

KATHLEEN WOODWARD

In the third volume of his expansive trilogy on the American experience, Daniel Boorstin tells a small story about a nineteenth-century inventor which so perfectly illustrates the intimate connection between technology and American culture—or more precisely, our fiction of it, the American dream—that we almost suspect it of having been fabricated by our popular imagination. The story springs from the 1846 tragedy of the Donner party, a group of pioneers trapped in the snowy Sierra Nevada as they made their way from east to west, and were forced to eat their own dead to survive. This horrific event in American history has inspired artists to explore the double tragedy of death and survival (I am thinking especially of Herbert Blau's theater piece, *The Donner Party: Its Crossing*), but that is the minority report, not Boorstin's story: he tells us instead about Gail Borden's pragmatic response, so flatly American in character, to mortality.

Moved by the suffering of the Donner party, Borden devised a practical solution to what he doggedly perceived as only a problem rather than a testament to the human condition. Problems, by definition, can be solved and thus catastrophes re-interpreted as mere difficulties. Technology provides the means. Borden, an enterprising *bricoleur* with a playful imagination, determined to find a method to make food more portable. In 1849 he discovered "an improved process of preserving the nutritious properties of meat, or animal flesh, of any kind, by obtaining the concentrated extract of it, and combining it with flour or vegetable meal, and drying or baking the mixture in an oven, in the form of a biscuit or cracker." This in turn led to the invention of condensed milk for which his name became a household word. Alleviating misery and creating community through technology—this is one aspect of the American experience. We see it portrayed jubilantly in Whitman's work. As Thomas Reed West has commented on Whitman's "Passage to India," Whitman envisions the "engineering labors of the modern world as representing a march toward the unity of the earth: the binding of regions and peoples through the canal, the railroad, and the cable. . . ."

But, as the acting out of Borden's story reveals, there is a dialectic to such dreamwork. The simple material process of homogenization—the technology of preserving milk—has implications for the whole of culture. Technological innovation triggers cultural change: as we have seen, standardized packaging diminishes diversity and has created twentieth-century global communities which are united only by consumption patterns. Borden's unqualified ardor for the technological process of condensation, applied to all areas, is worse than impetuous, it is adolescent. And we see it renewed, with equally cheerful naiveté, one hundred years later in the technological utopianism of Buckminster Fuller. Condense, and embrace future shock. "Condense our sermons," Borden advised a minister, "the world is changing. In the direction of condensing. . . . Even lovers write no poetry, nor any other stuff and nonsense, now. They condense all they have to say, I suppose, into a kiss." Progress is defined in technological terms, meaning is compressed into easily digestible units. As Borden continues, "I mean to put a potato into a pillbox, a pumpkin into a tablespoon, the biggest sort of watermelon into a saucer. . . . The Turks made acres of roses into attar of roses. . . . I intend to make attar of everything." And indeed the popular culture industry has followed this path: *Reader's Digest* packages intellectual pablum. From noble intentions, the American dream, spurred on by commercial interests, has been reduced to something of everyday middle class proportions: the single family dwelling.

Let me repeat: the story of Gail Borden is a small one, not a parable. The origins and history of the American dream are of course much more complex than this. If, from the perspective of Europe, the New Land was imagined as an Eden of earthly beauty and riches, it was also perceived as a retreat from political and religious persecution, a haven where new social ideals of democracy—economic, political, and social equality—could be put into practice and individual liberty realized. I do not mean to deny this vision of the American idea, nor do I need to point out here the contradictions of our history. But as so many of our social and economic historians have pointed out, our obsessive emphasis on abundance has displaced, and not so gradually at that, the original, more meaningful articulation of the American dream. We are, as David Potter has so lucidly argued, a people of plenty. Today our collective belief in material progress, or desire for it, dominates all others. Exchange value is the yardstick by which we measure our national, and often personal, satisfaction. But this is not a recent phenomenon. The desire for prosperity was central to the American character from the beginning. Thomas Jefferson, in his 1801 inaugural address, extolled "a rising nation, spread over a wise and fruitful land, traversing all the seas with the rich productions of their industry." "With all these blessings," he asked, "what more is necessary to make us a happy and prosperous people?" More than a century later Warren Harding campaigned on a platform of "Prosper America first." And by the 1950s the American dream could be summed up in shopping list style as "college enrollment, automobile ownership, foreign travel, longevity, the 'paperback boom,' home ownership, [and] social mobility." If abundance and natural resources were closely related in the eighteenth century, in the nineteenth and twentieth centuries, the emphasis shifted to *abundance achieved by technology*. In a pioneering essay on "Technology and Democracy, 1800–1860," the historian Hugo Meier illumines this relationship. He writes:

> If the utilitarian propensities of a democratic environment proved favorable to technological development as Tocqueville described it in the 1830s, no less real was the relationship of that technology after 1800 to the democratic goal of improving the lot of the common run of man. Indeed, the very utilitarian base of technology made it an instrument especially useful for furthering the ideal of "the greatest good for the greatest number." Since to the American "good" was in many ways becoming synonymous with "goods" (as Tocqueville so frequently emphasized), the goal of mass production became a conscious obligation imposed upon science and invention.

But the optimistic relationship between technological growth and democratic objectives which Meier traces in the years between 1800 and 1860 was subverted, if indeed it ever really existed, by the end of the century. As Lewis Mumford shows in *The Pentagon of Power,* the increasing domination of the technological impulse in the New World was responsible for the increasing concentration and perversion of power. The promise of social equality, personal autonomy, and mutual aid vanished. The pentagon of power has usurped, in Mumford's phrase, "the prerogatives of nature and the promises of utopia."

If technology has been a dominant concern of the American character, how has the American literary mind responded to its threats and promises? David Madden has asserted that "it is certainly true that the voice of most serious American literature since the Industrial Revolution got rolling has been a thunderous NO to the machine as a fabricator of American Dreams." Agreed. In the twentieth century we see this in novels ranging from Upton Sinclair's *The Jungle,* in which industrial technology dehumanizes the work process, and Joseph Heller's *Catch-22,* in which the technology of social engineering extinguishes human autonomy, to Thomas Pynchon's *Gravity's Rainbow,* in which multinationals and missile age technology spawn madness and masturbatory fantasy. In the latter, the American dream has turned into an international nightmare. Catastrophe is understood to be the bottom line of the human condition, and postindustrial technology—and its social organization—one of the conditions of that predicament. Although such literature as Robert Pirsig's *Zen and the Art of Motorcycle Maintenance* and much science fiction militate against Madden's view, *generally speaking* the American literary imagination of the twentieth century—representatives, we might say, of the humanities—have adopted an anti-technological stance. But how do we evaluate this basic response? How do we qualify it? What method do we use to do this research?

The basic model for such study is Leo Marx's *The Machine in the Garden.* What an intelligent, lucid book it is. Marx examines the American experience of technology through the interaction of American literature and the larger domain of American culture. He moves beyond a strictly literary scrutiny of his material into what is appropriately called American studies. He finds that although the American literary dream is that of a pastoral "retreat to an oasis of harmony and joy," nonetheless the overwhelming American response to the spectacle of the machine in the nineteenth century was "exhilarating." Further, Marx places his analysis in an arena larger than that of American studies. He invokes Freud, whose *Civilization and Its Discontents* taught us that the making of culture itself (in this case, industrial culture) imposes stress which we must inevitably seek to escape. Thus Marx concludes that the urge to withdraw from industrial culture is not really a specific reaction to a unique historical situation but an instance of a general phenomenon inherent in the dynamics of cultural change itself. The metaphors may change—the intrusion of the machine into the garden is an invention, an elaboration, an insight—but the structure of the metaphor remains constant. This theoretical framework lifts Marx's research out of an historically concrete, possibly parochial realm and allows him the delight of reading American literature through the eyes of Virgil and Shakespeare. But it also denies particularity to his study of the relationship between a certain kind of technology at a certain point in time and the American literary imagination.

The four essays which make up this section do focus directly on the relationship between technology and American culture at specific points in time. To do this work, all the authors find it necessary, as did Marx, to borrow insights, texts, and methods from other disciplines. The result is research which is rich and far-reaching.

We are fortunate to open a volume which is devoted primarily to essays by literary critics with one drawn from the typewriter of an historian of technology. Carroll Pursell's "The American Ideal of a Democratic Technology" provides us an historical context in which to read American literature. Pursell also avoids a pitfall common to many who study technology: he does not discuss *technology* per se but rather is careful to distinguish between *technologies.* Using Lewis Mumford's categories of democratic and authoritarian technics, Pursell traces what he sees as the unhealthy growth of authoritarian technics from the Republic's origins to the Second World War. While the development of an authoritarian technics has fulfilled, for many, the material promises of the American dream, he argues that it has also choked the roots of liberty. But, he notes, fortunately, the ideal of a democratic technics—characterized by simplicity, flexibility, and cheapness—has been rediscovered in the sixties and today goes by the name "appropriate technology." I should also add that Pursell does not perceive his topic as a mere academic one. For him scholarship and right action go hand in hand.

Joseph Slade's essay on "Cybernetic Discontinuity" turns to the American literary treatment of technology during the first four decades of the twentieth century and finds it weak. Slade judges the literature of this period, marked by Twain on the one hand and Faulkner on the other, in terms of the adequacy of its response to a particular technological change: the gradual shift from what he calls the First Industrial Revolution to the Second Industrial, or Cybernetic, Revolution. If the threat to the American dream in the nineteenth century was the machine which invaded the garden, in the early twentieth it was systems which had swollen out of control. Slade indicts American letters for its uncritical hostility toward technology and its refusal to recognize that man makes technology in his own image. He writes: "To the writer, whose office it was to shape the imaginative templates for experience, should have fallen the responsibility for creating metaphors that would help Americans see that behind the perversion and trivialization of invention lie the extensions of man himself." Thus, for Slade, technology, if properly understood, if not confronted with the inadequate knee jerk antagonism of the "bewildered" humanist, can indeed help us achieve worthy objectives. Slade moves beyond the texts of literature to the fascinating history of technology in order to unravel the reasons for the inarticulateness of literature, which he calls, following William Ogburn's theory of cultural lag, the "literary lag."

The third essay in this section—Miles Orvell's "Reproduction and 'The Real Thing': The Anxiety of Realism in the Age of Photography"—turns to the influence, we might well call it, of a technological process, photography, on literary production in America. Orvell's analysis calls into question David Madden's assertion that American literature is congenitally anti-technological. Showing how the uneasy relationship between the American literary imagination and realism (particularly photographic realism) has changed over time, from Hawthorne to Barthelme, from the American Renaissance to postmodernism and intermedia, Orvell reveals literature's shifting ambivalence toward, acceptance of, and indifference to its technological models. Whereas Hawthorne's "Birthmark" warns us not to attempt to rival nature (I might note here that in Poe and Hawthorne we find an American equivalent of Mary Shelley's *Frankenstein*, a nineteenth-century prophecy of twentieth-century genetic engineering), and the literature of the twenties, taking its cue from Stieglitz's work in photography, paradoxically restores us to a sense of the actual in an era of mass reproduction, much contemporary American writing, Orvell points out, literally incorporates the stuff of popular culture with pop art fascination but an edgier spirit. The fight between the two mediums, Orvell contends, has been called off: the contemporary realist no longer feels the need to compete with photography or the mass media. I should add that one of the fundamental dimensions of Orvell's essay is that it asks us, implicitly, to consider the differing, odd, often philosophically perplexing meanings of production and reproduction, presentation and representation. What is the difference between production and reproduction? Walter Benjamin's important essay comes to mind: "Art in the Age of Mechanical Reproduction." What a marvelously strange phenomenon Benjamin calls attention to—*mechanical reproduction*. For the root meaning of reproduction is of course biological, organic, but our technology allows us to imitate this process mechanically. Orvell's essay asks: what does it mean to reproduce the real thing? what is the relationship between interpretation and reproduction?

The last essay, by Michel Benamou, sees things differently and from a different perspective. Skillfully, sensitively edited by Charles Caramello after Benamou's death from his work in progress, the essay maps contemporary attitudes (we might more accurately call them fictions) about technology as they are projected in the discourse about our technological culture and its alternative futures. Benamou returns us to the belligerent tone of controversy: those who promote the planetarization of American technological culture—Buckminster Fuller and Marshall McLuhan, for example—pit themselves against those who challenge it. What characterizes postindustrial culture, Benamou asks, and just what is the relationship between the discourse about technology and the increasing technologization of our contemporary literature? Once again we return to the question of power with which Pursell opens this section on "Technology and American Culture." If the central issue is the distribution of power in a society, then our concern must be the relation of literature to that system of power. Benamou puts it this way: "How can the literary mind transform its power from a choice of words into a choice of world?"

The American Ideal of a Democratic Technology

CARROLL PURSELL

In 1815 Thomas Jefferson, asked to comment on a steam engine developed by a fellow American, compared it to those he had seen in Europe, and articulated the American ideal of a democratic technology scaled to human needs:

> I see, indeed, in yours, the valuable properties of simplicity, cheapness and accomodation to the small and numerous calls of life, and the calculations of its power appear sound and correct. . . . The importance of your construction will be enhanced by the consideration that a smaller engine, applicable to our daily concerns, is infinitely more valuable than the greatest which can be used only for great objects. For these interest the few alone, the former the many. I once had an idea that it might perhaps be possible to economize the steam of a common pot, kept boiling on the kitchen fire until its accumulation should be sufficient to give a stroke, and although the strokes might not be rapid, there would be enough of them in the day to raise from an adjacent well the water necessary for daily use; to wash the linen, knead the bread, beat the hominy, churn the butter, turn the spit, and do all other household offices which require only a regular mechanical motion. The unproductive hands now necessarily employed in these, might then increase the produce of our fields. . . . Of how much more value would this be to ordinary life than Watts and Bolton's thirty pair of mill-stones to be turned by one engine. . . .[1]

[1]Thomas Jefferson, Letter to George Fleming, 29 December 1815, *The Writings of Thomas Jefferson*, ed. H. A. Washington (Washington: Taylor and Maury, 1854), II, pp. 504–05, 507.

Jefferson's remarks serve to remind us that throughout American history there has been a continuing (though never dominant) realization that, as Lewis Mumford phrased it, "those who are concerned with maintaining democratic institutions" must see to it that their "constructive efforts . . . include technology itself."[2] Never triumphant, but never completely disappearing, this same desire for a technology scaled to human needs and aptitudes has reappeared in recent times in the movement for what is called Appropriate Technology. Both the dream and its antithetical nightmare have best been described, I think, by Mumford—in the whole body of his work, but most succinctly in his 1964 essay titled "Authoritarian and Democratic Technics." "My thesis," he wrote, "is that from late neolithic times in the Near East, right down to our own day, two technologies have recurrently existed side by side: one authoritarian, the other democratic, the first system-centered, immensely powerful, but inherently unstable, the other man-centered, relatively weak, but resourceful and durable."[3]

Authoritarian technics, wrote Mumford, was centralized and "drew on inventions and scientific discoveries of a high order: the written record, mathematics and astronomy, irrigation and canalization; above all, it created complex human machines composed of specialized, standardized, replaceable, interdependent parts—the work army, the military army, the bureaucracy." From the beginning these "work armies and military armies raised the ceiling of human achievement" in both construction and destruction. In our time these technics have provided the masses with an astonishing array and abundance of material goods, but, as Mumford warns, "by progressively fulfilling this part of the democratic promise, our system has achieved a hold over the whole community that threatens to wipe out every other vestige of democracy."[4]

Opposed to this is a democratic technics, which, in Mumford's words, is characterized by "the small scale method of production, resting mainly on human skill and animal energy but always, even when employing machines, remaining under the active direction of the craftsman or the farmer, each group developing its own gifts, through appropriate arts and social ceremonies, as well as making discrete use of the gifts of nature."[5]

The conflict between these two technics has waxed and waned throughout recorded history, but its modern intensification began at the same time as our United States. Brooke Hindle has called the simultaneity

[2]Lewis Mumford, "Authoritarian and Democratic Technics," *Technology and Culture,* 5 (Winter 1964), 7.

[3]*Ibid,* p. 2.

[4]*Ibid,* pp. 3, 6.

[5]*Ibid,* pp. 2–3.

of the American and Industrial Revolutions "one of the staggering coincidences of American history."[6] Mumford is more apprehensive than excited by the juxtaposition: "At the very moment Western nations threw off the ancient regime of absolute government, operating under a once-divine king," he wrote, "they were restoring this same system in a far more effective form in their technology, reintroducing coercions of a military character no less strict in the organization of a factory. . . ."[7] The resulting conflict had two facets, both of which have drawn attention throughout our history. In order to insure the survival of democracy one must preserve a democratic technics, but to do this, first, one must keep authoritarian technics from concentrating in the hands of the few, and second, one must at the same time provide easy access to a democratic technology for the many.

The prospect was hopeful during the years of the Early Republic and, as Hugo Meier and others have pointed out, it soon became a major canon of our American faith that democracy and technology were symbiotic, that political, social, and economic liberty would cause technology to flourish, and that technology in turn would help defend and extend such liberty.[8] Asserting in 1792 that "*free* governments are the only *asylum* and nursery for *science* and the *arts*," Joseph Barnes explained that:

> Free governments have always been found most disposed to encourage the rise and progress of science, and the arts; for this obvious reason, that in republican governments *merit* alone is, or *ought* to be the *standard* for character; which necessarily *excites* that *laudable* spirit of emulation, so essential to society; and which never fails to produce celebrated philosophers, statesmen, husbandmen and artists; the necessary effect of which, is, not only, the promotion of science and useful arts, but, the *happiness* of men,—the primary, the grand *object* of their existence.

Summarizing his argument, he maintained that "*Philosophy* being the *source* of liberty—*liberty*, of all the *improvements* in science and the arts; the

[6]Brooke Hindle, *Technology in Early America* (Chapel Hill: Univ. of North Carolina Press, 1966), pp. 17–18.

[7]Mumford, p. 4.

[8]Hugo Meier, "Technology and Democracy, 1800–1860," *Mississippi Valley Historical Review*, 43 (March 1957), 618–40. For other efforts to tie technology to American thought and character see Hugo A. Meier, "American Technology and the Nineteenth-Century World," *American Quarterly*, 10 (Summer 1958), 116–30; John E. Sawyer, "The Social Basis of the American System of Manufacturing," *Journal of Economic History*, 14 (Fall 1954), 361–79; Morrell Heald, "Technology in American Culture," *Stetson University Bulletin*, 62, No. 3 (Oct. 1962), 1–18; and Marvin Fisher, "The Iconology of Industrialism, 1830–60," *American Quarterly*, 13 (Fall 1961), 347–64.

improvement in science and the arts, of all the *real accomplishments* of the human species; consequently, *philosophy* is *essentially* the source of all virtue. . . ."[9] Benjamin Franklin's gesture with regard to his Pennsylvania Fire-Place exemplifies this faith in the interdependence between philosophy, or democracy, and technology. Instead of accepting a patent proffered by the Governor of his colony, he advertised it in a pamphlet complete with drawings, description, and instructions to masons for setting it up. He declined the patent, he said, "from a Principle which has ever weigh'd with me on such Occasions, viz. That as we enjoy great Advantages from the Invention of others, we should be glad of an Opportunity to serve others by any Invention of ours, and this we should do freely and generously."[10]

Other factors fostered a democratic technics during this period. The nation had recently been set up as a republic, not a monarchy, and a standing army had been rejected. Moreover, so long as technology relied upon the energy of sun and water, the triumph of authoritarian technics appeared unlikely, or at least avoidable. The potential energy of woods and farm land, falling water and blowing wind was widely and all but freely available, the technology to harness these sources of energy was cheap and widely known, and the sources themselves of such energy could not be monopolized. Even the coming of the American system of manufactures seemed to strengthen the democratic nature of American technology. In 1849 the touring Scottish chemist James F. W. Johnston visited an agricultural fair in Syracuse, New York, and noted that "the general character of the implements was economy in construction and in price, and the exhibition was large and interesting." "Ploughs, hay-rakes, forks, scythes, and cooking-stoves," he said, "were very abundant, and many of them well and beautifully made." The "potato grips and forks, of various kinds, cut from sheet-steel, were very elastic, light, strong, and cheap."[11] Only a few years later Joseph Whitworth and George Wallis wrote that "labour-saving machines are most successfully employed in the manufacture of agricultural implements. In a plough manufactory at Baltimore, eight machines are employed on the various parts of the woodwork. With these machines seven men are able to make the wooden parts of thirty ploughs per day." One result of the extension

[9]Joseph Barnes, *Treatise on the Justice, Policy, and Utility of Establishing an Effectual System for Promoting the Progress of Useful Arts, by Assuring Property in the Products of Genius . . .* (Philadelphia: Francis Bailey, 1792), pp. 9–10.

[10]Benjamin Franklin, *An Account of the Newly Invented Pennsylvanian Fire-Place* (Philadelphia: 1744), p. v. Quoted in Thomas R. Adams' introduction to the 1973 reprint (Boston: G. K. Hall and Co.).

[11]James F. W. Johnston, *Notes on North America: Agricultural, Economical, and Social* (Edinburgh: W. Blackwood & Sons, 1851), I, pp. 160–61.

of the American system from small arms to the manufacture of plows was that the implements made at his factory sold for from $2.50 to $7.[12] Surely the democratic nature of agriculture technology during the nineteenth century, and therefore its relatively low price, was as important as free land in extending a nation of yeoman farmers across a continent.

The American system, of course, with its machine designers and tenders, did not encompass the entirety of American technology. Many areas (farming and the building trades spring to mind) long resisted the intrusion of machine logic. This is one reason why many Americans were able (even when they were not forced) to move from job to job with easy proficiency. Whitworth and Wallis noted that "the citizen of the United States knows that matters are different with him [than with workers in Britain], and seems really to pride himself in not remaining over long at any particular occupation, and being able to turn his hand to some dozen different pursuits in the course of his life."[13] Karl Marx at one point quoted a Frenchman who had moved to San Francisco: "I was firmly convinced that I was fit for nothing but letterpress printing. . . . Once in the midst of this world of adventurers who change their occupation as often as they do their shirt, egad, I did as the others. As mining did not turn out remunerative enough, I left for the town where in succession I became a typographer, slater, plumber, etc. In consequence of thus finding out that I am fit for any sort of work I felt less of a mollusk and more of a man."[14]

Not content to leave well enough alone, however, Americans wanted to progress, and progress was away from this democratic technic. Early in the century Oliver Evans had confidently asserted that "we see by daily experience, that every art may be improved."[15] Late in the century, when Mark Twain's Hank Morgan gained power in King Arthur's Court he recorded that "the very first official thing I did, in my administration—and it was on the very first day of it, too—was to start a patent office; for I knew that a country without a patent office and good patent laws was just a crab, and couldn't travel any way but sideways and backwards." "The first thing you want in a new country," he concluded, "is a patent office; then work up your school system; and after that, out with your newspaper."[16] Little wonder that an anonymous scribbler was

[12]Joseph Whitworth and George Wallis, *The Industry of the United States in Machinery, Manufactures, and Useful and Ornamental Arts* (London: n.p., 1854), pp. 19–20.

[13]*Ibid*, p. vi.

[14]Quoted in Daniel Bell, *Work and Its Discontents* (New York: League for Industrial Democracy, 1970), p. 43fn.

[15]Oliver Evans, *The Young Mill-wright and Miller's Guide* (Octoraro, Pa.: Francis Bailey, 1807), p. 355.

[16]Mark Twain, *A Connecticut Yankee in King Arthur's Court* (New York: Signet, 1963), pp. 58–59.

inspired by an industrial exhibition in 1828 to write these lines:[17]

Genius of Art! What achievements are thine?—
  To delight and astonish the mind
The efforts of knowledge and fancy combine;
  But Triumph to thee is assigned.
How high swell the bosoms of patriots with pride,
To behold the rich treasures which here are supplied;
  While Invention's bright wand
  In the artisan's hand,
Points to glory the freemen of Freedom's own land.

But authoritarian technics were not so easily ignored. The power fantasies of such inventors and social engineers as J. A. Etzler had a democratic appeal but an authoritarian promise. In 1841 Etzler published a sequel to his book *Paradise* in which he attempted to prove, "from experience:—How to cultivate 20,000 acres by one machine and three or four men, with a capital of less than one dollar per acre, in the most superior mode—how to clear land from trees, stumps, roots and stones; fill and drain swamps, make dams, canals, ditches, roads, and perform any kind of work in the ground; build houses, and furnish as much inanimate power as desired, for ever, for any place and any stationary machine—all by the same system." In an appeal that was to be echoed many times over the next century in America, Etzler besought: "Ye, who are poor and of the labouring class, do not be alarmed at this invention. It will not deprive you of your poor pittance for sustaining life, obtainable now by your petty drudgery. When 20,000 acres of garden can be cultivated by three or four men, with but one dollar capital (once every year) per acre, you may easily become share-holders of joint stock companies, by paying from twenty to thirty dollars, and even this small sum, partly or entirely, in work of from one to three months, once for ever, and then enjoy the produce of from ten to twenty acres for each share, without further trouble or expense—and then live like gentlemen and ladies, to much better purposes than those of a dull animal in a tread mill."[18] It is surely significant that talk of mechanical slaves increased precisely at the time human slavery was increasingly condemned. The extension of modern technology to agriculture came not only through the introduction of labor-saving machines, but also through the rebirth of the work armies of the ancient Near East. In their recent book on slavery, *Time on the Cross,* Robert Fogel and Stanley Engerman point out that the

[17]"An Ode," 1828, published at an industrial exhibition. Copy kindly provided by Bruce Sinclair, University of Toronto.

[18]John Adolphus Etzler, *The New World: Or, Mechanical System to Perform the Labours of Man and Beast by Inanimate Powers, That Cost Nothing for Producing and Preparing the Substances of Life . . .* (Philadelphia: C. F. Stollmeyer, 1841), p. 4.

regimen of the large southern plantation was "more like a modern assembly line than . . . the routine in many factories in the ante-bellum era." Indeed they go so far as to assert that "the great plantations were the first large, scientifically managed business enterprises," and that black slaves were "the first group of workers to be trained in the work rhythms which later became characteristic of industrial society."[19]

The giant railroad companies of the period were also sources of the growing authoritarian technics, rationalizing for their own needs and convenience everything from business forms to political corruption and the establishment of local time. Significantly, railroads were among the earliest corporations to make systematic use of industrial scientists. Just as significant, for many Americans, was the fact that such technics were welcomed despite their anti-democratic tendencies. Speaking in the aggregates necessary to authoritarian technics, the astronomer and political economist Simon Newcomb insisted first that "the question whether the effect of any policy is good or bad depends very largely upon whether it increases or diminishes the sum total of the products necessary for human welfare," and then that "carrying passengers forty miles an hour for two or three cents a mile is a fact which outweighs all we can say about watered stocks."[20]

Thus, in one field of American endeavor after another, what Mumford called "the machine and the mechanical collective" triumphed over democratic technics.[21] The arts of war and self-government were revolutionized. Agriculture, the backbone of democratic technics, was industrialized, mechanized, made energy-intensive and ever-more efficient. Farmers who could not compete with the new technological system of machines, irrigation, and fertilizers fled to the cities and became dependent upon those who remained behind. These, in their turn, however prosperous, were locked into a tightening circle of technological restraints.[22] But nowhere was the process more obvious and dramatic than in the area of transportation. The balanced systems of a previous era were replaced by a primary reliance on one machine, the automobile. In its production first man-the-maker was rationalized through the technics of Fordism, then man-the-consumer was rationalized through the

[19]Quoted in C. Vann Woodward, rev. of *Time on the Cross: The Economics of American Negro Slavery*, by Robert Fogel and Stanley Engerman, *New York Review of Books*, 2 May 1974, 5–6.

[20]Simon Newcomb, *A Plain Man's Talk on the Labor Question* (New York: Harper and Bros., 1886), pp. 28, 61.

[21]Mumford, p. 6.

[22]For two excellent short articles on the recent history of American agribusiness, and the food it produces, see Jim Hightower, "Hard Tomatoes, Hard Times: Failure of the Land Grant College Complex," *Society*, 10 (Nov.-Dec. 1972), 10–11, 14, and 16–22; and Daniel Zwerdling, "Death for Dinner," *New York Review of Books*, 21 Feb. 1974, 22–24.

technics of Sloanism.[23] With transportation, as with food and most other of life's needs, Americans were asked—or forced—to accept efficient, increased production with its machine logic and rhythms. For economic growth had always been seen in America as a substitute for social justice. As Frederick Winslow Taylor told a congressional committee in 1912, under the benign rule of his new scientific management "both sides take their eyes off the division of the surplus as the all-important matter, and together turn their attention toward increasing the size of the surplus until this surplus becomes so large that it is unnecessary to quarrel over how it shall be divided."[24] It was, in short, what Mumford has called the "magnificent bribe." By accepting the new technics, and by living "under the democratic-authoritarian social compact, each member of the community may claim every material advantage. . . . But on one condition: that he must not merely ask for nothing that the system does not provide, but likewise agree to take everything offered, duly processed and fabricated, homogenized and equalized, in the precise quantities that the system, rather than the person, requires."[25]

Not everyone, of course, acquiesced willingly in this compact. In 1896 a Minnesota state senator wrote disgustedly to Ignatius Donnelly:

> While we think, brag of it, how far we are ahead of any former civilization, I for one am disgusted of the bragging and the boasting and simply believe it is not true. . . .
>
> I have heard it asserted that the printing Press, telegraph etc. have educated the masses, that the direful relapse will not come again as in the past. . . .
>
> Bosh! Our would be masters have a corner on the whole outfit of the inventions, and they are now just as much employed to the destruction of human rights as formerly, in the absence of those inventions, the peoples ignorance was used as a means.[26]

But this corporate and political monopolization of the very technology which was supposed to preserve the people's liberty, as recently described by David Noble in his book *America by Design*,[27] was one of the problems which motivated those who had some vision of a democratic

[23] I take the terms Fordism and Sloanism from the excellent book by Emma Rothschild, *Paradise Lost: The Decline of the Auto-Industrial Age* (New York: Random House, 1973). See also James J. Flink, "Three Stages of American Automobile Consciousness," *American Quarterly*, 24 (Oct. 1972), 451–73.

[24] "Taylor's Testimony Before the Special House Committee" (1912), in Frederick Winslow Taylor, *Scientific Management* (New York: Harper & Bros., 1947), pp. 29–30.

[25] Mumford, p. 6.

[26] Quoted in Norman Pollack, *The Populist Response to Industrial America* (New York: Norton, 1966), p. 22.

[27] David F. Noble, *America by Design: Science, Technology, and the Rise of Corporate Capitalism* (New York: Alfred A. Knopf, 1977).

technics. One promising area of reform was thought to be in electricity. By 1915 engineers were confident they could transmit economical amounts of electricity up to one hundred miles from central stations—far enough to provide nearly all Americans with adequate, flexible, cheap, clean energy. The fact that as late as 1935 less than ten percent of farms were so served, and the fact that rates in the United States were substantially higher than, for example, in Canada, led some critics to see in a reform of the electrical utility industry a new opportunity to fulfill the dream of a democratic technics. In the 1920s Morris L. Cooke, the Progressive engineer, produced for Governor Pinchot of Pennsylvania a comprehensive Giant Power plan which, without making power generation public, would have accomplished the twin goals of cheap and abundant available energy.[28] After a limited start in the Tennessee Valley Authority (1933), the Rural Electrification Administration, set up in 1935 with Cooke as its first director, finally succeeded in pushing rural lines to nearly all American farms.[29] Paradoxically, the coming of electricity tended as often as not merely to tie the individual farmer more closely into spreading national networks of communication and the agribusiness industry. As James Carey and John Quirk have pointed out in their article "The Mythos of the Electronic Revolution," the progressive "faith that electricity will exorcise social disorder and environmental disruption, eliminate political conflict and personal alienation, and restore ecological balance and a communion of man with nature," has unfortunately proved as baseless as the nineteenth-century hope that iron and steam would, as Timothy Walker put it in 1831, "perform all the drudgery of man, while he is to look on in self-complacent ease."[30] One fault was that while electricity might supply the clean, abundant, cheap, and divisible power needed for a democratic technology, it was provided by an increasingly authoritarian technics, the private power utilities. And although during the Great Depression many democratizing measures were introduced into the Congress, which sought to break the monopoly that big business (including private power utilities) had on research talent and technical expertise of the nation, none of them were adopted.[31]

[28]See Jean Christie, "Giant Power: A Progressive Proposal of the Nineteen-Twenties," *Pennsylvania Magazine of History and Biography*, 96 (Oct. 1972), 480–507. For a later episode in Cooke's career, see Jean Christie, "The Mississippi Valley Committee: Conservation and Planning in the Early New Deal," *The Historian*, 32 (May 1970), 449–69.

[29]See the official history of the REA, *Rural Lines—USA: The Story of the Rural Electrification Administration's First Twenty-Five Years, 1935–1960*, USDA Misc. Pub. No. 811 (Washington, D.C.: GPO, 1960), p. 36.

[30]James W. Carey and John J. Quirk, "The Mythos of the Electronic Revolution," *The American Scholar*, 39 (Spring 1970), 219–41 and (Summer 1970), 395–424; Timothy Walker, "Defense of Mechanical Philosophy," *North American Review*, 23 (July 1832), 122–36.

[31]See for example H. R. 1536, 75th Cong., 1 sess., introduced by Jennings Randolph: "A

When the nation began its defense mobilization in 1940, that effort followed the lines of the status quo rather than the possibilities inherent in a world war for democracy.

Led by Vannevar Bush, the nation's spokesmen for science and technology deliberately chose to follow two cardinal principles in gearing up for war: first, the Axis was to be defeated as soon as possible; and second, insofar as possible, the mobilization was to be carried on with no disruption of the domestic status quo. What this last commitment meant was that since most research and development were concentrated, during the prewar period, in private industry—and within private industry within a very few large firms—the war effort was to be carried out largely in terms advantageous to the most powerful institutions in America. Thus, wartime spending for new science and technology was channeled largely through such corporations as General Motors, AT&T, and General Electric, and through such educational institutions as Harvard, M.I.T., and the University of California.[32]

Such an elitist mobilization was not inevitable. Indeed, hearings before the Kilgore committee during the war witnessed a steady stream of protestors who charged that American technology was being underutilized, and that research and development funds were being unnecessarily concentrated to make the nation's rich even richer.[33] Critics such as I. F. Stone and Bruce Catton pointed out that a democratic war effort was the best guarantee of a democratic postwar society.[34] The resulting legislation suggested by the Kilgore committee would have drafted patents, laboratories, and research brains, but, as in the previous decade, those who stood to benefit from the growing centralized technics were strong enough to bury such reformist proposals. On the eve of the war Waldemar Kaempffert, science editor of *The New York Times*, worried that "mass consumption, mass recreation, mass distribution of energy, and the collectivist utilization of identical things are impossible without control of mass production, without organization. The inventors have standardized behavior, pleasures, tastes. There is less freedom than there

---

Bill to aid and promote scientific research of a basic character upon which the inception and development of new industries or the expansion of established industries may be dependent, to encourage increased effort on the part of individuals toward the further advancement of scientific knowledge and discovery and for other purposes."

[32]See Carroll Pursell, "Alternative American Science Policies during World War II," in *World War II: An Account of its Documents*, ed. James E. O'Neill and Robert W. Krauskopf (Washington, D.C.: Howard Univ. Press, 1976), pp. 151–62.

[33]One of Kilgore's pieces of legislation was S. 2721, 77th Cong. 2 sess. For the text, and some of the hearings, see James L. Penick, Jr., et al., *The Politics of American Science: 1939 to The Present*, 2nd ed. (Cambridge, Mass.: MIT Press, 1972), pp. 82–95.

[34]I. F. Stone, *Business as Usual: The First Year of Defense* (New York: Modern Age Books, 1941); Bruce Catton, *War Lords of Washington* (New York: Harcourt, Brace & Co., 1948).

was a century ago because of invention; there will be still less to-morrow."[35] When the war was over, the stage was set for a corporate intensification of these trends.

At the same time, the decade of the late fifties and early sixties saw the rediscovery of a host of problems which Americans had thought long-since solved. Betty Friedan's book *The Feminine Mystique* (1963) demonstrated that women's rights had not been secured totally by the right to vote. Ralph Nader's *Unsafe at Any Speed* (1965) attacked the most American of machines and recalled the continuing abuses of corporate responsibility. Michael Harrington reminded the nation of the existence of *The Other America* (1962), that third of the nation which still lived in poverty. And Rachel Carson, in her book *Silent Spring* (1962), proved that the Progressive conservation crusade had not solved all of the nation's environmental problems. This rediscovery and revitalization of the ideal of a democratic technics rested on no one book (though several struck the same theme), but rather grew with the resurgence of environmental concern, the development of the so-called counterculture, and with the successive revelations of technological arrogance in such diverse but critical places as Detroit, Vietnam, and the board rooms of the petroleum industry. In 1967 the late Paul Goodman mourned the "betrayal of the promise of independent scientific technology" envisioned by the idealists of the nineteenth century. "They thought," he wrote, "of science as humble, brave, and austere, and of technology as circumspect, neat, and serviceable. Working by its own morale, scientific technology should by now have simplified life rather than complicated it, emptied the environment rather than cluttered it, and educated an inventive and skillful generation rather than a conformist and inept one."[36] In a later essay Goodman called upon technicians to fulfill their proper roles as moral philosophers, acting with foresight, caution, and prudence.[37]

Many of those who today advocate a technology characterized by foresight, caution, and prudence, are to be found working under the banner of Appropriate (or Alternative or Intermediate) Technology. Popularly identified with the late British economist E. F. Schumacher and his slogan, "Small is Beautiful," the movement is often focused on Third World nations. VITA, for example, the Volunteers in Technical Assistance, founded in 1959, advises developing countries on the "use of skills

[35]Waldemar Kaempffert, *Science Today and Tomorrow* (New York: Viking, 1939), pp. 261–62.

[36]Paul Goodman, "The Morality of Scientific Technology," *Dissent*, 14 (Jan.-Feb. 1967), 43.

[37]Paul Goodman, "Can Technology Be Humane?" *New York Review of Books*, 20 Nov. 1969, 27–34. For some examples and problems, see Ralph Nader, et al., eds., *Whistle Blowing: The Report of the Conference on Professional Responsibility* (New York: Grossman Publishers, 1972).

and materials that are locally available, or if equipment has to be imported, on designs that can be repaired by the village craftsman."[38] Better known groups have concentrated on our own problems of *over-development*. Private efforts such as the New Alchemy Institute on Cape Cod and the Farallon Institute in California have gained some notoriety for their work. The founder of the latter group, Sim Van der Ryn, has recently been appointed State Architect of California by Governor Jerry Brown, himself an avowed disciple of Schumacher. Van der Ryn has established within his agency a new Office of Appropriate Technology which is attempting to influence other state agencies to adopt such technologies as solar power.[39] At the federal level, a newly funded National Center for Appropriate Technology is being established in Butte, Montana. In describing its effort, *Science* magazine gave as good a definition of the subject as we are likely to find: "Appropriate technology generally means technology that is small, easy to understand and maintain, cheap, dependent on local resources, and fitted to local needs. It makes heavy use of renewable resources . . . and makes minimal demands on capital and on nonrenewable environmental and energy resources."[40] Thomas Jefferson could hardly have put it better. Today the poet Gary Snyder echoes his concerns.

But this redefinition for our own time of the traditional ideal of a democratic technology answers only half the problem. There is a real danger that, as in the past, the technology that begins as democratic will be quickly absorbed into a larger authoritarian context. The history of energy use in America provides a case in point.

As mentioned above, during most of its first century, the United States depended upon widely available, easily harnessed energy sources: wind, water, wood (now called more gloriously "biomass"), and animal. They were, in terms of our definition, "small, easy to understand and maintain, cheap, dependent on local resources, and fitted to local needs." Since 1850, however, our consumption of energy has multiplied more than thirty times, and whereas wood supplied ninety percent of our needs in 1850, by 1900 coal had become the dominant fuel (providing seventy percent) and by 1950 oil and natural gas were the leading contributors (with 55.5 percent).[41] Not only had we become an energy-intensive society, but the source of that energy was increasingly non-

[38] *Science*, 188 (6 June 1975), 1000.

[39] For an interview with Van der Ryn see "What is Appropriate Technology?" *Cry California*, 12 (Summer 1977), 30–32.

[40] "NCAT: Appropriate Technology with a Mission," *Science*, 195 (4 March 1977), 857.

[41] Earl Cook, "The Flow of Energy in an Industrial Society," in *Energy and Power* (San Francisco: W. H. Freeman & Co., 1971), p. 85.

renewable and, as importantly, was increasingly vulnerable to monopolization.

Potentially, the emphasis of Appropriate Technology on solar and wind power to provide for our energy needs is a progressive step toward reestablishing the democratic nature of our energy mix. The potential, however, remains only that, and there is increasing evidence that it too might well become authoritarian. Ralph Nader has quipped that "if twenty years ago title to the sun had been given to Exxon, with a depletion allowance, we'd have solar energy today."[42] In effect, the government is today doing something much like that. As *Science* magazine has reported, "government planners seem to have trouble grasping" the point that solar power is "fundamentally different from other [dominant energy] sources. Solar energy is democratic. It falls on everyone and can be put to use by individuals and small groups of people." Instead, the government's "research program has emphasized large central stations to produce solar electricity in some distant future and has largely ignored small solar devices for producing on-site power."[43] It is obvious that on-site solar power would provide a democratic alternative to the energy monopolies, while plans for orbiting reflectors and "power towers" would enhance their position. As *Science* comments, "the centerpiece of the government's solar energy program is proceeding from small to large tests in a fashion that is remarkably parallel to the well-established pattern of nuclear reactor development. Big facilities, big expenditures, and a multidecade development program all characterize the program of centralized solar thermal generating stations."[44] Much the same story appears in the government's miniscule wind energy program.[45] Thus the fact is that this authoritarian pattern grows not out of the nature of the technology itself (solar and wind resources are by their very nature democratic) but rather from the larger social technics which prevail in this country.

Looking back over the course of American history, the role played by technology in bringing us to our present state is both obvious and too little studied. Scholars such as Hugo Meier have illuminated the predominant belief that American technology and democracy were not only compatible

[42]Quoted in *Santa Barbara News-Press,* 21 April 1974, p. B1, cols. 1 & 2.

[43]"Solar Energy Research: Making Solar After the Nuclear Model?" *Science,* 197 (15 July 1977), 241.

[44]*Ibid.*

[45]"Wind Energy: Large and Small Systems Competing," *Science,* 197 (2 Sept. 1977), 971–73; and "Windmills: The Resurrection of an Ancient Energy Technology," *Science,* 184 (7 June 1974), 1055–58.

but mutually dependent.[46] John Kouwenhoven has maintained that this democratic technology in fact defines the unique nature of American culture.[47] Leo Marx has brilliantly described and analyzed those voices of dissent which attempted to come to grips with both the pastoral and the industrial.[48] John Higham has suggested that by the mid-nineteenth century the United States had moved from a primordial, through an ideological, to a technological context for its national identity.[49] As we build upon these studies, however, it is important to keep several major points in mind. First, to an important degree, the American experience in this regard has been "advanced," but not entirely unique. Lewis Mumford has found the roots of authoritarian technics in the ancient Middle East and, in his monumental study of the *Myth of the Machine,* has traced the evolution of what he calls the pentagon of power.[50] Similarly, Lynn White has insisted, persuasively I think, that the roots of ecological crisis are not uniquely American, but extend back to fantasies of power originating in the Medieval period.[51] Second, it is important to remember that the voices of Jefferson and Thoreau, Mumford and Goodman, do not belong simply to an archaic and inert past, but are part of a *continuing tradition* of perception that technology does not simply affect social institutions, it *is* a social institution and, as such, is of proper concern to those who would preserve our democratic liberties. Third, it should also be obvious that technology is not indivisible, is not autonomous, and does not contain some inner logic which makes inevitable the kind of world in which we find ourselves. Indeed, the traditions of technology—simplicity, cheapness, and accommodation—are *those very virtues which Jefferson saw as characterizing a democratic technics.* That we have today a megamachine displaying exactly the opposite characteristics is not due to the logic of technology, but to some other force at work in American

[46]Meier, "Technology and Democracy, 1800–1860" and "American Technology and the Nineteenth-Century World" (see note 8).

[47]John A. Kouwenhoven, *The Arts in Modern American Civilization* (New York: Norton Library, 1967).

[48]Leo Marx, *The Machine in the Garden: Technology and the Pastoral Ideal in America* (New York: Oxford Univ. Press, 1964). See also his "American Institutions and Ecological Ideals," *Science,* 170 (27 Nov. 1970), 945–52. For British writers, see Herbert L. Sussman, *Victorians and the Machine: The Literary Response to Technology* (Cambridge, Mass.: Harvard Univ. Press, 1968).

[49]John Higham, "Hanging Together: Divergent Unities in American History," *Journal of American History,* 61 (June 1974), 5–28.

[50]Lewis Mumford, *The Myth of the Machine: Technics and Human Development* (New York: Harcourt, Brace & World, 1967) and *The Myth of the Machine: The Pentagon of Power* (New York: Harcourt, Brace, Jovanovich, 1970). Also see his "Apology to Henry Adams," *Virginia Quarterly Review,* 38 (Spring 1962), 196–217.

[51]Lynn White, Jr., "The Historical Roots of our Ecologic Crisis," *Science,* 155 (10 March 1967), 1203–07.

society. Finally, it is important to remember that this whole question is one of more than simple academic interest: the very survival of our democratic society is dependent upon an understanding of those forces working to destroy it. Authoritarian technics tend to an integration and uniformity which inevitably create a social and natural environment which is not only unpalatable but unstable as well. Openness and diversity, characteristics of a democratic technics, are essential for the kind of successful adaptions which our society must make to survive.

American Writers and American Inventions: Cybernetic Discontinuity in Pre-World War II Literature

JOSEPH W. SLADE

In 1906, in his autobiography, Henry Adams confessed to anxieties engendered by the social implications of the dynamo, the symbol, as he saw it, of a new century. Ten years later, in *his* autobiography, Henry's brother, Charles Francis Adams, Jr., admitted that his impulses toward reform as President of the Union Pacific Railroad had not been sufficient to deal with the social disruptions caused by the locomotive, the symbol, although he did not see it as such, of the century just past. Both men attributed their sense of powerlessness to the inability of the educated classes to cope with mechanization, but unlike his brother, Henry Adams made a brilliant attempt to understand the nature of his technological symbol. Of the two, the less ambitious response of Charles Francis Adams was the more typical of American writers during the first four decades of the twentieth century. Antipathy to technology, by and large covert in the nineteenth century, emerged openly in the next—aggravated by the seismic forces of economics, the standardization of industry, the urbanization of the countryside, the realignment of class structures, and the widening distance between science and letters. Deliberately or not, most writers assumed a posture of bewildered humanism, the essence of which was an inarticulateness regarding the machine. When they muckraked, they could be eloquent about the injustices of the factory system, and they became adept at uncovering the tentacles of octopi as industry expanded, but about the nature of the beast itself they were reticent.

When Vernon Parrington complained that the Gilded Age had produced few competent critics of industrialism among writers,[1] the observation might have been applied to the subsequent four decades as well. The inadequacy of response is noticeable chiefly in retrospect, for the First Industrial Revolution, which involved machines that do the work of human (or animal) muscles, was imperceptibly giving way to the Second, which involves machines that do the work of human senses and human intelligence. Strictly speaking, the Second Industrial Revolution commences with World War II as the result of advances in physics and electronics, but because technology develops more or less continuously, the roots of that revolution lie further back in time. During the early part of the twentieth century, inventions and their applications became steadily more complex and systematized, and systematization is one of the factors that distinguish the Second Industrial Revolution from the First.

For the most part, Americans then had difficulty swallowing the diversified extensions of technology, and more difficulty still in determining what they signified. American writers seemed conversant only with the simplest and most hackneyed of inventions. Still, when they did consider familiar machines, they were apt to treat them as strange devices and to emphasize the differences between men and inventions. In nineteenth-century letters, machines *invade* an American garden; in twentieth-century examples, systems *proliferate* out of control. The usual invader of the garden, the locomotive, was still awesome enough to continue generating railroad novels well into this century, and in the spread of that invention Robert Herrick would discover ominous presence, as in his description of a railroad in *Together* (1925) as "a vast organism, with a history, a life of its own, lying like a thick ganglia of nerves and blood vessels a third of the way across our broad continent, sucking its nourishment from thousands of miles of rich and populous territory." Herrick quite properly concentrated not on the locomotive, an invention already domesticated in the previous century, but on the systems that it had spawned. Despite its organic nature, entirely suitable to systems, his analogy is interesting principally because of the distance it establishes between man and technology, a characteristic common to native thought. There is no Frankensteinian tradition in our literature, and little recognition that machines are created in man's image. Were one to know the country's technology only through its literature, one might conclude that machines and systems were entirely alien to their creators.

[1]Vernon Louis Parrington, *The Beginnings of Critical Realism in America, 1860–1920* (New York: Harcourt, Brace and World, 1958), p. 137.

Because the machine was a product and an extension of man's head and hand, and because it had become so much a part of his everyday endeavors, it bore a special relationship to him. If the machine magnified man's fear, it also mirrored his aspirations. Clearly it benefited its creator, yet it also made demands upon him; take pride in it though man might, in its presence he felt impotent, anonymous, dehumanized. Had man not sensed the identification, had he not felt the intimacy with the machine, he would not have needed to erect alienation as a defense. It was difficult for the average citizen to put his apprehensions into words; it was no less so for the writer.

About this inarticulateness Robert L. Heilbroner has made a pertinent observation:

> When we ask ourselves exactly in what way the quality of life is adversely affected by the incursion of the machine, we find that however powerful the outcry against technology, the complaints are largely metaphorical. They describe, often very compellingly, the *feelings* of the observer, but they are rarely if ever explicit in describing the causal process leading to the end product they deplore.[2]

Writers, of course, deal in metaphors as a means of translating experience into literature. The problem—then as now—was not that writers employed metaphors nor even that they were hostile to technology, but that they stopped short of accepting metaphors explicit enough to explain their antipathy. To do so would have been to acknowledge the affinity between man and his mechanical creations, and that, apparently, was too disconcerting. Although the humanist typically masked his anxiety with concern for the machine's ugliness, vulgarity, or debasement of culture, the threat that actually disturbed was that technology at best would reduce man to a cipher, and at worst displace him altogether. When that prospect surfaced, as in Elmer Rice's *The Adding Machine* (1923), the most straightforward attack on technology in American literature and the only native work to compare in bluntness with foreign examples like Karel Capek's *R.U.R.* (1921) or E. M. Forster's *The Machine Stops* (1909), the result was near-hysteria.

Underlying the threat was a duality defined by historian Bruce Mazlish as "the fourth discontinuity," a term which can be modified—in recognition of the period's advances in automation and systematization—to cybernetic discontinuity. According to Mazlish, man has already resolved three discontinuities, or skewings of his organic relationship to his world, resulting from the Copernican, Darwinian, and Freudian

[2]Robert L. Heilbroner, "The Impact of Technology: The Historic Debate," *Automation and Technological Change,* ed. John T. Dunlop (Englewood Cliffs, New Jersey: Prentice Hall, 1963), p. 19.

revolutions, after each of which he had to muster again a belief in the continuum of nature. Mazlish calls the remaining discontinuity a dichotomy between humans and machines, and thinks that man has been struggling to eliminate it: "To put it bluntly, we are now coming to realize that man and the machines he creates are continuous and that the same conceptual schemes, for example, that help explain the workings of his brain also explain the workings of the 'thinking machine.' Man's pride, and his refusal to acknowledge this continuity, is the substratum upon which the distrust of technology and an industrialized society has been reared."[3]

Only one American writer of consequence explicitly "eliminated" the discontinuity, but Mark Twain did so for the wrong reasons, out of despair over man's benightedness or out of frustration over the Paige typesetter, on which he bestowed human traits. In *The Mysterious Stranger* (1916), and *What is Man?* (1906), especially the latter, Twain said flatly that man's brain was mechanical and man himself, a machine. This bridging of the discontinuity did credit to neither man nor machine, for it assumed that man had lost his humanity, although Twain knew that no machine was capable of the malice of his characters. Bitter as they were, Twain's remarks were not aimed at technology; he simply used machines as a negative reflection of man's indifference to his fellows. Lesser writers, stopping well short of Twain's pessimism, nevertheless believed the machine's influence to be deleterious. When they employed mechanical metaphors to describe humans or society, they suggested that the relationship between man and his creation was one-sided, with the machine nefariously dominant. More important, such writers implied that the comparison did violence to man's nature, that no human, unless he were depraved, resembled a machine. Men were becoming machines, said the writer, and society was becoming mechanical. But indifference, callousness, the diminution of originality and creativity—characteristics those metaphors were intended to convey—did not build machines in the first place. Writers seemed to have lost sight of the human in the inventor as well as in the invention. The effect, once again, was to depict technology as alien.

It was doubtless difficult for the writer to claim kinship with machines. It would also have been difficult for him to discover the causes for his feelings, if only because the effects of invention were so confusing. In 1916, the date of *The Mysterious Stranger,* a writer much younger than Twain could still have remembered the Civil War. He might have understood that the progress of certain types of inventions had led

[3]Bruce Mazlish, "The Fourth Discontinuity," *Technology and Culture,* ed. Melvin Kranzberg and William Davenport (New York: New American Library, 1972), pp. 217–19.

inexorably to the abolition of slavery in the United States, but he would have known as well that the perfection of other types had magnified the carnage of the conflict. Moreover, the technological horrors of World War I would shortly affront his sensibilities to an even greater extent. If he were aware that inventions were improving the lot of Americans, he could not have ignored the sweatshops that had exploited women and children until only a few years before. If he knew that inventions were reducing human hunger and insecurity, he could not be so certain that they were ensuring freedom and dignity. The very number and disparity of new inventions seemed to preclude rational assessment or compact Frankenstein-type metaphor. Even so, it should have been clear to the thoughtful writer that machines and systems were humanizing American culture at least to the degree that they dehumanized it. For their part, humanists should have been ready to direct not so much the "civilizing" of the machine, as John F. Kasson has described one phase of the evolution of American attitudes toward technology, as its humanizing. That did not require equating men and machines in Twain's terms, nor even abating distrust of invention; it did call for the affirmation of continuity. To the writer, whose office it was to shape imaginative templates for experience, should have fallen the responsibility for creating metaphors that would help Americans to see that behind the perversion and trivialization of invention lie the extensions of man himself.

Unable to conceive of machines as made in man's image, writers were just as incompetent as other Americans to fathom the problematic intimacy that the growth of technology thrust upon them. The misfortune was not so much that the writer had a special obligation to deal with the complexities of technology, though he did. Nor was it so much that his failure to devise adequate metaphors was a failure of humanism, though it was. What was unfortunate, however, was that the writer missed his chance to provide literary guidance in a critical period of American culture. By losing that role of leadership, by succumbing to the kind of helplessness that overwhelmed Charles Francis Adams, he fostered the birth of the zeitgeist theory of technology, which holds that man's inventions have a life of their own, beyond man's control, alien to his aspirations. Worse, to the extent that the trend prevailed, it rendered literary statement increasingly inconsequential in an increasingly technologial society.

There were, of course, exceptions. The writer who did most to humanize technology, and who discussed at length the metaphorical needs involved, succeeded in creating a symbol to rival Henry Adams's. In giving form to his symbol, Adams had drawn heavily on the scientific knowledge of his time, and from the breadth of his comprehension fashioned a monument of future-shocked abstraction. By invoking the Second Law of Thermodynamics, he concluded that society was in process of disintegration, the victim of energy gone entropic: multiplicity was destroying unity. In 1932 another American reared an abstraction upon a real monument, a fantastic bridge. Hart Crane's thesis was at odds with Adams's; as the poet envisaged it, technology wrested unity from diversity. Crane also arrived at his symbol through a method at variance with Adams's, or so he thought:

> For unless poetry can absorb the machine, i.e., acclimatize it as naturally and casually as trees, cattle, galleons, castles and all other human associations of the past, then poetry has failed of its full contemporary function. This process does not infer any program of lyrical pandering to the taste of those obsessed by the importance of machinery; nor does it essentially involve even the specific mention of a single mechanical contrivance. It demands, however, along with the traditional qualifications of the poet, an extraordinary capacity for surrender, at least temporarily, to the sensations of urban life. This presupposes, of course, that the poet possesses sufficient spontaneity and gusto to convert this experience into positive terms. Machinery will tend to lose its sensational glamour and appear in its true subsidiary order in human life as use and continual poetic allusion subdue its novelty. For, contrary to general prejudice, the wonderment experienced in watching nose dives is of less immediate creative promise to poetry than the familiar gesture of a motorist in the modest act of shifting gears. I mean to say that mere romantic speculation on the power and beauty of machinery keeps it at a continual remove; it cannot act creatively in our lives until, like the unconscious nervous responses of our bodies, its connotations emanate from within—forming as spontaneous a terminology of poetic reference as the bucolic world of pasture, plow and barn.[4]

Aside from some interesting contradictions—such as how one avoids the necessity of specifically mentioning "contrivances" while making "continuous poetic allusion" to them—the passage indicates that Crane understood the dimensions of technology. The "Pindar of the Machine Age," as Crane once styled himself,[5] built his Bridge upon a fetish for contrivances unequaled even in the novels of Lewis and Dos Passos; in "The Bridge" are celebrations of subways, acetylene torches, cars, airplanes, elevators, ships, forges, radios, locomotives, electric signboards. Certainly Crane did "surrender" to the urban environment, assimilated and transcended it, but in the doing also romanticized it and the technology which had transformed it.

[4]Hart Crane, "Modern Poetry," *The Complete Poems of Hart Crane,* ed. Waldo Frank (Garden City, N.Y.: Doubleday and Co., 1958), pp. 181–82.

[5]Quoted by Peter Viereck, "The Poet and the Machine Age," *Dream and Responsibility: Four Test Cases of the Tension Between Poetry and Society* (Washington, D.C.: Univ. Press of Washington, D.C., 1953), p. 52.

Alan Trachtenberg has already explored the technological motifs of Crane's splendid poem in *Brooklyn Bridge, Fact and Symbol* (1965). Trachtenberg demonstrates that Crane's belief that the Brooklyn Bridge had fulfilled American hopes by forging a symbolic "passage to India" was premature if not naive. At the same time, Crane's visionary romanticism, shade though it might into mysticism, acknowledged the triumphs of technology that a more conventional romanticism usually denied. Where Crane's contemporaries scorned the machine as a threat to individuality, Crane praised it as an agent of community; where they feared mechanical power out of control, he applauded its democratic vigor; where they rejected the materialism of technology, he embraced the spirituality of invention. More important, Crane understood that man and machine had merged, that humans and their technology penetrated each other: the American environment was not only urban, but also technological.

Although Crane believed that "use and continual poetic allusion" would "subdue [the machine's] novelty," it was the former rather than the latter factor that operated most strongly to diminish the excitement of technological innovations. Average Americans learned to live with and to use machines more quickly than did writers. The more Americans welcomed new machines, and they did so eagerly, the less writers were willing to accord the inventions serious literary attention. It was not that writers deliberately ignored machines, nor that they were necessarily suspicious of popular enthusiasms—nor was it that familiarity bred contempt. Rather it was as if, having admitted finally that the locomotive was here to stay, the literary mind was determined not to be shocked again. And, indeed, as inventions became more commonplace, assuming the status of items of furniture in home, office, and factory, the less sensational they seemed. On the one hand, the ubiquity of inventions ensured that references to them would enter poetry, fiction, and drama as Crane's "spontaneous terminology." On the other hand, the commonness of the new devices often precluded their being closely examined in a literary context. Awkward and mysterious, the telephone virtually escaped literary notice, even as it was being installed in thousands of homes, forever to alter the habits of millions of Americans.

In a sense, the machine's loss of glamor early in the twentieth century may have eroded its metaphorical potential for many writers; they may not have understood how sophisticated technology was becoming. Technological change is not always dramatic. A new invention does not necessarily supplant older ones. Radio, for example, did not replace newspapers and magazines, and has not itself been rendered obsolete by television. The effects of technology are largely cumulative, and for that reason can go unnoticed as they are assimilated by a culture. A writer attempting to apprehend technology imaginatively would have to be able to extract drama and human significance from a very gradual process. Henry Adams managed to do just that, of course, but the writers who followed him, it might be argued, did not have Adams's advantage. Before 1900, the most impressive advances in invention came in machines—like Adams's dynamo—designed as tools of production; after that date there appeared in increasing numbers technological devices intended for the personal benefit of the average citizen: telephones, gramophones, cinemas, vacuum cleaners, washing machines, refrigerators, radios. The automobile and the airplane should have been amazing enough, but both were products of lengthy, almost tedious development and on-going refinement. Between the two world wars stretches a span of years that Siegfried Giedion has designated the period of "full mechanization,"[6] when "mechanization takes command" by penetrating the most intimate areas of American life. Measured in dramatic achievements, i.e., capital tools like the dynamo, the pace of invention slowed, to be offset to a degree by that spread of consumer-oriented inventions that was to make the machine so familiar. Actually, then, while the between-the-wars era consolidated entrepreneurship and saturated the economy with certain types of products, it was also a phase of *technological lag* instead of acceleration that Henry Adams had predicted. The reasons were in large part social and economic rather than technological, according to Samuel Lilley, who identifies a faulty distribution system, mass unemployment, and monopoly as the three major factors which:

> tended in the years between the wars (and indeed from about 1900 onwards) to restrict the use of the most advanced methods and ultimately to slow down the progress of invention. First the faulty distribution system, which created the chronic difficulty of selling products and therefore reduced the incentive to introduce advanced techniques. Second, mass unemployment, which discouraged invention through the fear of displacing labour and, by depressing wage-rates, often made it cheaper to use monotonous hand labour instead of introducing highly automatic machines. Third, monopoly with its tendency to protect its own interests even at the cost of progress. These were not necessarily the only retarding factors at work, but at least they were three important ones.[7]

But the lag, less pronounced in the United States than in other countries, did not prevent the growth of automation, although it surely slowed it. It is perhaps too easy to point out benchmarks in the history of technology. For example, we commonly refer to the 1870s as important in

[6]Siegfried Giedion, *Mechanization Takes Command: A Contribution to Anonymous History* (New York: Norton and Co., 1969), p. 41.

[7]Samuel Lilley, *Men, Machines and History* (New York: International Publishers, 1965), p. 190.

the rise of American industrialism because there occurred then the massive shift from water power and iron to coal and steel, with deleterious consequences for milltowns, whose dark configurations mesmerized Victorian and modern writers alike. Or we might choose 1900 simply because Henry Adams saw his Dynamo in Paris that year, but such dates are largely matters of convenience. Nevertheless, the decades between the wars saw the transition of mechanics to cybernetics. In *Turning Points in Western Technology,* D. S. L. Cardwell observes that although self-controlling machines antedate this period, "one post-[World War I] trend was unmistakable: the development of the technology of control systems and the march towards increasingly automatic production processes."[8] Rudimentary automation techniques of that time prefigured the cybernetics systems of the present. To be sure, full cyberneticization had to wait for mathematical methods of automatic control and instrumentation like the semi-conductor and the computer, but the four elements of automation—mechanization, feedback, continuous process, and rationalization—were integrated between 1918 and 1939.

Shortly after Henry Adams published his reflections on his new goddess, Henry Ford geared his assembly line into continuous operation. Within a few more years, experimentation with hydraulic and pneumatic valves, and gradually, with electric sensors, established practical feedback systems. Americans began to note that rationalization—in the form of efficient reorganization of work—seemed to have as its object the reduction of humans to machine-like components within the systems. As Ford himself put it, in one of the pronouncements collectively known as "Fordism," the machines were rather more reliable than their symbiotic partners: "A business is men and machines united in the production of a commodity and both the men and the machines need repairs and replacements. . . . Machinery wears out and needs to be restored. Men grow uppish, lazy, or careless."[9] The exponents of Scientific Management, variously called Industrial Engineering or Taylorism, after its originator, Frederic W. Taylor, were more explicit. For them, all human effort could be redesigned with stopwatches and motion studies.[10] Rationalization, the efficiency experts said, aimed at converting a worker into "an interchangeable part of an interchangeable machine making

[8]D. S. L. Cardwell, *Turning Points in Western Technology* (New York: Science History Publication, 1972), p. 204.

[9]Quoted by Emma Rothschild, *Paradise Lost: The Decline of the Auto-Industrial Age* (New York: Vintage Books, 1973), p. 57.

[10]Taylorism precipitates a textile factory revolt in Sherwood Anderson's *Perhaps Women* (1931).

interchangeable parts."[11] Beginning around 1910, wide application of the principles of Scientific Management to industry resulted frequently in the kind of worker disaffection that a later generation would call alienation.

And yet, however repugnant the identification of man and machine to writers, occasionally, from surprising sources, came the recognition that the popular metaphor had positive connotations. In 1911, for example, Samuel Gompers assessed the metaphor in *The American Federationist:*

> Let science, the intellectuals, the employers and the body politic take hold of the laborer from his earliest years and build him up, clear up to what the original man-maker put into him in the way of physical and mental possibilities. Let the promoters of the proposition study how the laborer may be given his full physical height and breadth, his full mental and moral growth, his full potential worth to society as a good machine. In the end they might find that they had contributed in the construction of a good strong man.[12]

Gompers was less interested in ideology than in economic participation in a technological society for his followers. Still, so astonishing a concession on the part of a labor leader indicates how powerful was the tendency to equate the worker and the tool. In his own way, Gompers had found what Peter Drucker has called "the essential flaw" in the science developed by Taylor and his disciples: "The worker is put to use as a poorly designed, one purpose machine, but repetition and uniformity are two qualities in which human beings are weakest."[13] The challenge to industry, then and now, has been to design systems that utilize men as efficiently as machines, but that task requires an understanding of similarities and differences between them.

More intimate with machines than were writers, and more prone to ascribe virtues to their tools, the workers Gompers led were not distressed by their chieftain's metaphorical flight. The famous Hawthorne experiment of the 1920s demonstrated that Western Electric workers took pride in their machine-like efficiency—when they were asked for their opinions and allowed to set their own pace—when they were permitted, in short, to participate in factory life like the superior machines that they were. Forced by necessity to perceive that men were similar to machines, and aware that machines performed more and more of their crude functions, such workers needed such metaphors to

[11] Quoted by Melvin Kranzberg and Joseph Gies, *By the Sweat of Thy Brow: Work in the Western World* (New York: Capricorn Books, 1975), p. 160.

[12] Quoted by David R. Weimer, "The Man With the Hoe and the Good Machine," *Studies in American Culture: Dominant Ideas and Images,* ed. Joseph J. Kwiat and Mary C. Turpie (Minneapolis: Univ. of Minnesota Press, 1960), p. 72.

[13] Quoted by Kranzberg and Gies, pp. 162–63.

conceptualize their relationship to their assembly-line environment. And, as the quotation from Gompers indicates, those metaphors came not from mainstream literature but from other sectors of society.

If experience in the factory were not enough to establish the resemblance between man and machine, ideas current during the period reinforced it. Books like Carl Snyder's *The World Machine* (1911), Jacques Loeb's *The Mechanistic Conception of Life* (1912), or George Crill's *Man, An Adaptive Mechanism* (1916) expanded the analogy: not only was man himself a machine, but he lived within one, either because nature herself was mechanistic or because industrialism had made his environment artificial. Most such works drew on the clockwork models inherent in still-viable Newtonian physics, or attempted to adapt, however misguidedly, the thought of men like Darwin, Freud, and Marx. Mass production, with its emphasis on the integration of man and machine, seemed compatible with the ideas of Darwin and the corollaries of Spencer and Durkheim. Freud's division of the psyche into forces, drives, and regressions suggested a mechanistic structure of the personality. Marx, of course, had understood that factories and industrial institutions, whatever else they might be, were forms of social organization. Moreover, Marx believed that man's relationship to the things he produces was the most significant factor in human society, and further, that the individual was related to other humans primarily through his production. The implication was that man and thing are interchangeable: if the production and distribution of things were modeled on social relationships, then the equation seemed to be reversible, so that men were things—or, if one likes, machines.

Americans did not have to accept Marxian analysis to see that advancing technology was wrenching capitalism and creating new classes capable of manipulating new social structures. During this period Joseph Schumpeter and lesser economists, as much aware of the power of corporations as muckrakers like Steffens and Tarbell, began speculating on the bureaucracies of Standard Oil, American Tobacco, Quaker Oats, Otis Elevator, and Singer Sewing Machine. During this period also occurred the rise of the "managerial elites" and "the revolt of the engineers,"[14] movements that generated others like "The New Machine" of Henry Gantt and the "Technocracy" of Howard Scott, which enjoyed vogues of some duration. Fundamental to such organizations was the notion that, given large systems and sophisticated machinery, and given the affinity between man and machine, machine experts or engineers should run not only the factories but society as well, to unify human endeavor under order and efficiency.

[14]See Edwin T. Layton, *The Revolt of the Engineers: Social Responsibility and the American Engineering Profession* (Cleveland: Case Western Reserve Univ. Press, 1971).

Advocates of The New Machine and Technocracy were long on hope and short on theory, but the prevalence of the three connected assumptions—that man resembled a machine, that his environment had already to a degree been mechanized (or systematized), and that technological experts should govern society—gave impetus to a series of books which have been called "technological utopias." Between 1883 (John Macnie's *The Diothas; Or a Look Ahead*) and 1933 (Harold Loeb's *Life in a Technocracy: What It Might Be Like*), at least twenty-five Americans published these visions of technological grandeur. In virtually all of them, according to Howard Segal, the principal authority on the subject,

> the envisioned domestication of technology and nature alike is what leads to the resolution of the tension Leo Marx deems irresolvable: the tension between the industrial and the agrarian orders, the tension between the machine and the garden, a tension which Marx believes lies at the heart of "the American experience." That tension is to be resolved by the modernization rather than the abandonment of the garden, by its being transported out of the wilderness and relocated in the city, but a city now itself transformed from a lethal chaos to a healthy order. The new "industrialized garden" does not take the form of Marx's "middle landscape": the Jeffersonian ideal of an agrarian-based yet technologically proficient yeoman republic. Rather, it takes the form of what in the "real" world have since been termed "megalopolises": massive combinations of urban and suburban tracts embracing practically all of, in this case, utopia.[15]

Such schemes—some of which have materialized in the present, though without the blessings envisioned by the utopians—were an approach to resolution of the man-machine discontinuity, for they were predicated on the assumption that people could live harmoniously with their inventions. Despite their visionary aspects, these sub-literary blueprints managed to avoid the simplistic industrial-pastoral tension that continued to disturb more established writers.

That tension weakened as the spreading presence of machines and systems gradually relegated the pastoral ideal to the status of remote rebuke to the urban areas of the twentieth century. It no longer mattered so much whether the Garden was invaded by the machine or left untouched; rural America no longer served as an escape valve for urban pressures. That was the lesson of "the revolt from the village," the literary recognition that agrarian cultures imposed a conformity and a regimentation as great as the factory's, and spawned *boobus Americanus* as readily as metropolises dumped Maggies on the streets. As might be expected, the industrial-pastoral tension survived longest in those

[15]Howard P. Segal, "American Visions of Technological Utopia, 1883–1933," *The Markham Review*, 7 (Summer 1978), 65–76.

sections whose citizens prided themselves on their distance from the industrialized East. Perhaps the best-known testimonial to the persistence of the agrarian mythos was offered by Twelve Southerners in *I'll Take My Stand*. The protests against technology in that volume seemed antiquated even in 1930, especially when considered in the light of the Country Life Movement of a few years earlier, a program that did much to ensure the viability of declining rural areas by providing trained county agents, machinery, and technological assistance to farmers beset by ignorance and poverty.

Writers in the West also remained hostile to alterations wrought by technology on nature. The ironies could be intense, for where the Garden still flourished, as California Boosters boasted it did in that state, it blossomed through the grace of technology that irrigated and fertilized it, harvested it, and transported its bounty along vast networks of trade. From laments on the passing of geographical, social, and psychic frontiers can be extracted two by Californians which illustrate disapproving perceptions of the technology which was reshaping the territory. The first is George Sterling, who, conscious of the encroachments of man's invention on his "unspoiled" Carmel, asked, "Am I to be but a whisper amid thunders, a flower insecurely rooted in the concrete of crowded ways, a dwarf dazed among the enginry of his own devising?"[16] The second, better known, is from Robinson Jeffers's "The Purse-Seine":

> I thought, We have geared the machines and locked all together into interdependence: we have built the great cities; now
> There is no escape. We have gathered vast populations incapable of free survival, insulated
> From the strong earth, each person in himself helpless, on all dependent. The circle is closed, and the net
> Is being hauled in. They hardly feel the cords drawing, yet they shine already. The inevitable mass-disasters
> Will not come in our time nor in our children's, but we and our children
> Must watch the net draw narrower, government take all powers—or revolution, and the new government
> Take more than all, add to kept bodies kept souls—or anarchy, the mass-disasters.
>
> These things are Progress. . . .[17]

Despite apparent similarities, the examples exhibit differences attributable to talent and orientation. Sterling was a *fin de siècle* poet; his flower in the dust metaphor derived from nineteenth-century individualism, and

[16]George Sterling, untitled manuscript, c. 1920, Sterling Collection, Mills College.

[17]Robinson Jeffers, "The Purse-Seine," *The Modern Age*, ed. Leonard Lief and James F. Light (New York: Holt, Rinehart and Winston, 1969), pp. 443–44.

his protest was a simple variation on the industrial-pastoral dichotomy. Although both men understood that man is a prisoner of his own ingenuity, Jeffers, a poet of more modern bent and greater ability, apprehended also a dominant characteristic of twentieth-century technology—that it unites as it permeates culture and in so doing redefines the relationship of the individual to society. Jeffers's insight into the systematization of the landscape was remarkable for 1937, especially when it is set against the hackneyed vision of "satanic mills" and "industrial-hells" that are the stock indictments of technology among essayists of the twenties and thirties. Jeffers's analogy is an improvement on Herrick's organic one because it stresses the man-made dimensions of the technological dilemma. Sophistication like Jeffers's was rare, for his criticism, entirely valid and still relevant, grew out of his perception that the advance of technology involved more than the trampling of flowers. His nightmare vision of impotent interdependence was the dark side of Crane's dream of technologically-achieved community. The industrial-pastoral tension thus did not disappear—it is with us to this day—but as it faded the hostility to invention that it had stimulated was matched by other literary reactions occasionally just as sharp.

The antipathy of some writers reflected a repudiation of modernity itself, or a rebellious opposition to traditional American practicality, or just a romantic preference for an instinctual rather than a rational approach to life. Sometimes the hostility was snide, the result of a distaste for the machine's vulgarity. Into this category falls Henry James's sniff at "the great religion of Elevator" and the indignities of riding in one of Mr. Otis's contraptions in *The American Scene* (1907).[18] Sometimes it was superficial, the concomitant of a preoccupation with the more absurd effects of rampant "progress." Inveigh as he did against Fordism, chain stores, and the mores of an increasingly homogenized society, H. L. Mencken could say as late as 1931 that "We live in a Machine Age, but there are still plenty of us who have but little to do with machines, and find in that little no answer to our aspiration."[19] By then Walter Lippmann had for a decade been attacking "the pseudo-environment" created by information systems which filtered ideas and affected the behavior of Americans, but in Mencken's defense we should remember the technological lag. Sometimes the hostility was anachronistic, the effect of ignorance. Amused by Willa Cather's belief that horses on a farm were superior to thrashers, Anthony Hilfer remarks that "There is a case to be made against a machine civilization, but Miss Cather lacked the intellec-

[18]Henry James, *The American Scene* (New York: Harper and Bros., 1907), pp. 180–81.

[19]H. L. Mencken, *The Vintage Mencken*, ed. Alistair Cooke (New York: Vintage Books, 1958), p. 202.

tual ability to make it."[20] Occasionally the scorn was churlish and silly, as when Waldo Frank, appalled at the crowds lining up in front of the Roller Coaster at Coney Island, sputtered: "No person in whom joy is overflowing will resort to a machine that stands him on his head and spins him on his ear, to shake it out. These are makeshifts of despair. America is a joyless land."[21]

The better writer was far more ambivalent. His ambivalence was not much different from that of the average citizen, proud of and grateful for his new automobile and his wife's sewing machine, but alienated by his work down at the shop. Nor was it more remarkable than the traditional American ambivalence toward education. Not to acknowledge that machines could liberate as well as enslave would have been similar to denying that education could improve the quality of life. Moreover, inventing was almost an American pastime, and so was playing with inventions, as Sinclair Lewis's Babbitt toyed with his car and his electric cigar lighter. Respect for "know-how" made folk heroes of Henry Ford and Thomas Edison as readily as it had Eli Whitney and Samuel Colt for earlier generations. Eddie Rickenbacker, a car racing champion at sixteen, an air ace at twenty-eight, had become a success by following childhood advice to hitch his wagon to a machine. Charles A. Lindbergh had written *We* (1927) to celebrate his fusion with the Spirit of St. Louis, and had gone on to invent (with Alexis Carrel) a perfusion pump called an artificial heart. The more marvelous the machine, the more contiguous with man's mind or his body, the more easily it could be exalted. Worship of the purity, the discipline, and the power of machines provided a tempting escape from the confusion of shifting values.

The wish to worship furnished Eugene O'Neill with a subject for *Dynamo* (1929), a particularly appropriate choice given the primacy of that symbol for Henry Adams. The protagonist of the play is not dehumanized by machines; he has already been dehumanized by a Fundamentalist minister-father and a jealous, ignorant mother. Electricity represents liberation for him, and onto the dynamo that generates the current he projects his fears and guilt and desire to comprehend his world:

> But there must be a center around which all [life] moves, musn't there? There is in everything else! And that center must be the Great Mother of Eternal Life, Electricity, and Dynamo is her Divine Image on earth! Her power houses are the new churches! She wants some one man to love her purely and when she finds him worthy she will love him and give him the secret of truth and he will become the new savior who will bring happiness and peace to man.[22]

[20] Anthony C. Hilfer, *The Revolt from the Village, 1915–1930* (Chapel Hill: Univ. of North Carolina Press, 1969), p. 99.

[21] Waldo Frank, *The New America* (London: Jonathan Cape, 1922), p. 192.

[22] Eugene O'Neill, "Dynamo," *The Plays of Eugene O'Neill* (New York: Random House, 1954), III, 479.

Whether one considers this feminization of electricity an inevitable avatar of Henry Adams's Virgin-Dynamo, or just—in Leo Marx's phrase—"the technological sublime" gone berserk, O'Neill makes clear that this god is as false as the one the boy's father worships. At first glance the American readiness to personify the machine, reminiscent of Emerson's desire to anthropomorphize the locomotive,[23] seemed to speak to the man-machine discontinuity, but the real urge, O'Neill believed, was to deify. The play attacks the slavish faith in technology that O'Neill saw Americans drifting toward: the machine was still an alien presence, not a creation of man's own devising, still a danger to Americans uncertain of technology's role in American society. That O'Neill could discuss his subject sympathetically indicates, however, that major writers of the four decades before World War II could bring some understanding to that role.

During this period the number of writers of repute who were uncritically in favor of technology reduces—if we leave out Hart Crane, whose response was complex—to one, Carl Sandburg. What is surprising, however, is not that major authors did not share Sandburg's attitude, but that they so rarely lapsed into hostility of the kind we noted earlier. Sometimes the antipathy we would expect to find was forestalled by obsessions peculiar to an author. For example, despite the occasional glimpse of a Carrie Meeber at work before a shoe-mending device, the novels of Theodore Dreiser are too preoccupied with the blind forces of nature to waste time on anger at machines. Throughout Dreiser's fiction are scattered references to automatic machines taken for granted, the "usual" telephone switchboards, streetcars, and elevators; they make no great impact on the lives of his characters, save as the source of dull wonder.

Sometimes ambivalence grew out of the author's perception that machines were important to the common man. "The most interesting thing going on in American life is inside the factories," thought Sherwood Anderson, who wished to treat industrialism fictionally "to get the beauty and poetry of the machine, but at the same time its significance to labor."[24] Although he tried to do so in books like *Marching Men* (1917), *Poor White* (1921), and *Perhaps Women* (1931), the efforts were not memorable, perhaps because Anderson could not heed Hart Crane's warning that "mere romantic speculation on the power and beauty of machinery keeps it at a continual remove" from human endeavor. Perhaps also Anderson's own unsettling experience in a paint factory worked against his intentions.

[23]See John F. Kasson, *Civilizing the Machine: Technology and Republican Values in America, 1776–1900* (New York: Penguin Books, 1977), pp. 120–21, 126, for an account of Emerson's tendency to anthropomorphize the machine.

[24]Sherwood Anderson, *The Letters of Sherwood Anderson*, ed. H. M. Jones and W. B. Rideout (Boston: Little, Brown, 1973), pp. 210–11.

Usually ambivalence developed from a passive, unconscious recognition that machines and systems were firmly entrenched in the twentieth century. If such writers were not hostile to technology, it is in part because they did not focus on it. Having become part of the environment, inventions and their extensions served chiefly as background for narratives. F. Scott Fitzgerald was one of the first Americans to remark that the automobile had enlivened the sexual activity of the young;[25] he was surely aware that the radio had broadcast the rhythms of the Jazz Age, and he understood that the cinema had heightened the expectations and redefined the lifestyles of his contemporaries. For all that, he does not deal with inventions as such; they are just *there.* When Fitzgerald does refer to them specifically, inventions serve as icons of class: automobiles as big as the Ritz, airplanes that span the coast for those rich enough to ride in them, tickertapes that click out money and privilege.

Ernest Hemingway's themes similarly obscure the evolving significance of machines in human contexts. In Hemingway's war novels, complex weapons destroy civilians and shred traditional values. In some of his sports stories his heroes disdain mechanical aids and opt for primitive techniques. Yet in other works cars, boats, and planes function simply as extensions of man; they are tools whose morality is determined by their use. This commonsense attitude makes it possible on occasion for Hemingway to view technological devices as intimate elements of the lives of his men and women and as appropriate details of his landscapes. In *The Sun Also Rises,* the wheezing bus that carries Jake Barnes, Bill Gordon, and a flock of Basques up mountain roads to Burgette receives affectionate treatment in the narrative because it inspires humor, stimulates camaraderie, and gets the two Americans within striking distance of an unspoiled trout stream. The vehicle is not alien to the setting. The terrain it traverses, however, is European. When we compare this scene with the edenic, machineless American Garden of "Big Two-Hearted River," Hemingway's position is far from clear. Alliances between man and machine in his fiction are uneasy.

The uneasiness is more pronounced in the work of William Faulkner, who is also more thoughtful in his ambivalence toward technology than Hemingway. Had Faulkner's novels not been so grounded in the rural South, and were his ambivalence not so keyed to this region, he might eventually have addressed the man-machine discontinuity directly. At any rate, complex machines for Faulkner are emblematic of the twentieth century, whose collisions with the nineteenth-century world of Yoknapatawpha County wear down his Mississippians. Faulkner's early novels are cautionary tales of the danger of empty metaphor and futile resistance

[25] F. Scott Fitzgerald, *The Crack-Up,* ed. Edmund Wilson (New York: New Directions, 1956), p. 14.

to change. In *Sartoris,* Southerners who fume and sputter against the automobile shrink beside the figure of their ancestor, Colonel Sartoris, who mastered the locomotive and built a railroad. More specifically, when Faulkner allows his own love of inventions to surface in books like *Pylon,* he can describe the leader of a barnstorming aviation troupe as "single-purposed, fatally and grimly without any trace of introversion or any ability to objectify or ratiocinate, as though like the engine, the machine for which he apparently existed he functioned, moved, only in the vapor of gasoline and the filmstick of oil."[26] Faulkner thus aligns himself with Samuel Gompers in admitting that man and machine do resemble each other; the necessary qualification is that man is the more complex of the two—he is not, or should not be, "single-purposed."

Nevertheless, far from indicating any serious attempt to resolve the man-machine discontinuity, the ambivalence of the finest literary minds represents a sidestepping of the issue. The factor which best explains why Anderson, Fitzgerald, Hemingway, and Faulkner were not hostile to machines also explains why they could not confront the discontinuity and why they could not advance beyond popular thinking on the subject of technology. Put simply, the reason is that these men did not envision machines as in themselves inimical to human society. For them, mechanical devices and mechanical processes were less troubling than the forces of twentieth-century commercialism. In *The Sound and the Fury,* for instance, Faulkner has Jason Compson use the money he swindled from Caddy to buy an automobile even though Jason is nauseated by the smell of gasoline. The machine is not a symbol of Jason's evil, nor is Jason like the machine. His sterility, his rapaciousness, and his alienation condemn him, and these traits Faulkner associates with commercialism.

It is true that these writers refer occasionally to the "mechanization" of life in "the machine age." Such lapses may be attributed to confusion and to the residual distrust that the most practical of humanists still feels before the most innocuous of machines. Like the average American, thoughtful writers identified inventions with progress, and knew that progress was real. If life did seem mechanized, it was sometimes easier for writers to follow the inclinations of the man in the street and saddle business interests with the blame. In addition to being a popular reaction, moreover, the approach had intellectual respectability. The distinction between inventions and the commercialism that perverted them had been established early in the century by one of America's most idiosyncratic minds, albeit a non-literary one.

The two American thinkers most directly concerned with technology before World War II were Lewis Mumford and Thorstein Veblen. In his

[26]William Faulkner, *Pylon* (New York: Signet Books, 1962), pp. 102–03.

volumes on architecture and in works like *Technics and Civilization*, Mumford discussed the stages of mechanization and the elegance, the discipline, and the symbology of the machine. Guiding his investigations was the hope that the organic potentialities of systems could be made harmonious with man's needs. Although the influence of Mumford's erudition should have been considerable, it was Veblen's voice that carried in the literary arena, largely because of the praise heaped on *The Theory of the Leisure Class* (1899) by William Dean Howells. In his later years, Veblen would provide ammunition for the Technocrats, but prior to the twenties he devoted little attention to the social utility of engineers. Beginning with *The Theory of Business Enterprise* (1904) and continuing with other volumes, Veblen implicitly eliminated the cybernetic discontinuity.

Where the Marxists had accepted the institutionalist role of technology, Veblen asserted that technology was a natural outgrowth of rationalism. Convinced that technological development was a dynamic, causal factor behind advancing civilization, he insisted that Americans had to come to terms with their inventions. Technology was the flowering of man's most creative instinct. Moreover, because of the reciprocal relationship between man and machine, proliferating technology could only further stimulate creative, i.e., rational, habits and impulses in humans. Thus, although the man on the assembly line might seem to be coarsened by his experience, he was actually learning discipline and rationality; given sufficient time the worker would become the "good machine" Samuel Gompers desired—but without the qualities Gompers deemed essential. *The Theory of Business Enterprise* is replete with passages in praise of the machine's influence: "The machine throws out anthropomorphic habits of thought. . . . It inculcates thinking in terms of opaque, impersonal cause and effect. . . . Thus in the nature of the case the cultural growth dominated by the machine industry is of a skeptical, matter-of-fact complexion, materialistic, unmoral, unpatriotic, undevout."[27] The more humans emulated the machine they themselves built, the better off they would be.

But if machines were good, and if to be like machines were good, how did it happen that technology was so often perverted? The answer, said Veblen, was that opposed to man's creative instinct was another he called "predatory," which resided in its purest form in the American businessman. "The production of goods is a mechanical process, incidental to the making of money," Veblen wrote in 1919, "whereas the making of money is a pecuniary operation, carried on by bargain and sale,

[27]Thorstein Veblen, *The Theory of Business Enterprise* (New York: Charles Scribner's Sons, 1904), p. 313.

not by mechanical appliances and powers. . . .[28] The fault lay not with the machine, but with its economic corruption, not with the inventor but with the entrepreneur. The upshot was that Veblen eliminated the man-machine discontinuity by saying that the machine was vested with the best human characteristics; but in the process he replaced one duality with another, the dichotomy between industry and business. The result, however, was to deflect fear of the machine onto the businessman.

The problem with this deflection is that while it permits the critic of technology to inveigh against a satisfactory enemy it leaves unanswered the more basic question of man's relationship to his inventions. The man-machine discontinuity does not lose its tension, and the metaphorical difficulties remain, as the work of Sinclair Lewis illustrates. Because of Veblen's emphasis on rationality as the most human of traits, the acceptance of his theories was hardly wholesale among writers, but Lewis did endorse the general outline. As Mark Schorer has observed, Lewis's novels popularized Veblen's ideas: "Looking back again at *Babbitt* from this point of view, we can see it as a dramatization of the divorce between industry and business, of the disappearance in a transitional America of independence and individuality in the machinelike processes of mass culture."[29] Like other Americans, Babbitt suffers from metaphorical contradictions. On the one hand, he can perceive his "way of life as incredibly mechanical. Mechanical business—a brisk selling of badly built houses. Mechanical religion—a dry, hard church, shut off from the real life of the streets, inhumanly respectable as a top-hat. Mechanical golf and dinner-parties and bridge and conversation. Save with Paul Riesling, mechanical friendships—back slapping and jocular, never daring to essay the test of quietness."[30] On the other hand, he feels best when he visualizes himself as mechanical: "He felt superior and powerful, like a shuttle of polished steel darting in a vast machine."[31] Babbitt's aesthetic sense is most stimulated by machinery, although Lewis undercuts his response with irony: "He had enormous and poetic admiration, though very little understanding, of all mechanical devices. They were his symbols of truth and beauty. Regarding each new intricate mechanism—metal lathe, two jet carburator, machine gun, oxyacetylene welder—he learned one good realistic-sounding phrase, and used it over and over, with a delightful feeling of being technical and initiated."[32]

Veblen believed that literary hostility to technology was fed by

[28]Thorstein Veblen, *The Vested Interests and the Common Man* (New York: Viking Press, 1946), pp. 91–92.

[29]Mark Schorer, *Sinclair Lewis: An American Life* (New York: McGraw Hill, 1961), p. 772.

[30]Sinclair Lewis, *Babbitt* (1922; rpt. New York: New American Library, 1961), p. 190.

[31]*Ibid*, p. 45.

[32]*Ibid*, p. 58.

snobbery, a suspicion not without foundation. His stringent humanism did not erase that hostility, for it still flourishes. Nevertheless, to the degree that writers could translate Veblen's opposition between technology and business into the traditional schism between art and commerce, they muted their distrust of the machine. This warming trend was signalled most clearly with the publication of John Dos Passos's *Manhattan Transfer* (1925), which established the machine as a permanent facet of American culture, and his three-volume *U.S.A.* (1930, 1932, 1936), which established it as something more. In the later works the technologist stands on a par with the artist, as much a benefactor of the quality of life as any creator, and subject to the same economic pressures. As countless critics have noted, the rhythms of *U.S.A.* are the rhythms of the machine, the pulse of a modern society in motion. Dos Passos's characters take their humanism where they find it, and they discover it at least as often in the presence of the machine as they do in its absence. By employing the "camera eye" and the "newsreel" and by pacing his narrative with biographies of Taylor, Ford, Steinmets, Veblen, Edison, and others, Dos Passos achieves that "spontaneous terminology" Hart Crane believed would guarantee the absorption of technology into literature, and he also makes that terminology work metaphorically.

Whether one considers it an adequate response to technology or not, Dos Passos's work was soon dated by the feverish wartime engineering that revved up the Second Industrial Revolution in earnest. World War Two's computers, rockets, and radars, which in their turn would be succeeded by microprocessors, smart bombs, and lasers, would seem more alien than ever to writers, who had never made their peace with earlier inventions. From war's end on, social, political, and economic imperatives would remove the prerogative of metaphor from literature, and vest it instead in popular culture or in non-literary disciplines. The moviemaker like Kubrick, the cartoonist like Feiffer, the political scientist like Ferkiss, the sociologist like Bell, even the odd literary critic like McLuhan, seemed better equipped than the writer to conceptualize man's relationship to his inventions. Today the writer's task is to synthesize other people's metaphors. So far, only Thomas Pynchon has successfully accepted that challenge, and it is significant that he began his career with a parody of Henry Adams's Virgin. Should other writers follow Pynchon's lead, they may regain the ability to speak of American inventions. Since the age of cybernetics has arrived, it is time to span the man-machine dichotomy, and time to end the literary lag.

Reproduction and "The Real Thing": The Anxiety of Realism in the Age of Photography

MILES ORVELL

In talking about the relationship between an artistic reproduction and the real thing of which it is an image, we are dealing necessarily with approximations. While physical objects can be mass-produced, one thing identical to another, artistic works cannot literally reproduce reality. Realism, as we normally talk about it, is a matter of conventions, of a rhetorical relationship between the author and reader by which the latter regards the image as "like reality." What is realistic for one era, however, is not realistic for another; and as Harry Levin has demonstrated, one age achieves its realism precisely by undermining the conventions of a preceding age. But however relative the term may be, the "confrontation of life by literature," to use Levin's phrase, is constant, and especially characteristic of the modern age.[1] I would add, though, that works of realism are not so much mirrors of life, as they are mirrors of what men regard at any time as "the real," which is a reflection of the philosophical and material environment they inhabit. And whatever the changes in the meaning of the term, my assumption is just the opposite of Harold Rosenberg's when he states that "realism, either esthetic or philosophical," is not "a characteristic of American thought."[2] One strong characteristic of American esthetic thought is, I think, precisely its

[1]Harry Levin, *The Gates of Horn* (New York: Oxford Univ. Press, 1966), p. 470.
[2]Harold Rosenberg, *The Anxious Object* (1964; rpt. New York: Collier, 1973), p. 78.

effort to come to terms with the real, its effort to rival nature and to rival the various technologies of reproduction such as painting and photography.[3] It is an effort that has been accompanied throughout by a certain anxiety over the mimetic process.

What is distinctive in the contemporary realist is that, while still seriously concerned with the process of seeing and describing the world, he has largely ceased to worry about the literal correspondence of image and real thing. If we go back to the nineteenth century and come forward again to the twentieth, we can see a gradual evolution in the anxiety of the realist, one that might be distinguished in three phases. The first phase covers much of the nineteenth century and is characterized by the artist's feeling that he must not be too literal in his creation. The second phase, a reversal of the first, runs from the teens through the 1960s, and is marked by the artist's feeling that he cannot be literal enough, that his model, in effect, is the camera. The third phase—the contemporary—dissolves the anxiety of the second, and, taking an entirely new attitude toward photography, ceases to worry about traditional questions of mimetic realism.

I am tracing what is in effect a subtheme in the history of realism, a reflexive dimension that comes to the fore at key points in the evolution of realism as a way of embodying the changing assumptions underlying the way the artist looks at the world. While the artist is normally concerned simply with creating reproductions of the real thing, according to the conventions of his time, occasionally he steps back and considers in his fiction the act itself of reproducing the real thing. The eye is so often the organ of perception for the writer as for the visual artist, that we should not be surprised that the visual artist—whether painter or photographer or filmmaker—should act as surrogate for the writer, or that the writer should reflect on his own art in terms of painting or photography. But there is a stronger case than mere analogy for considering the writer's concern with technologies of visual reproduction, for these technologies may profoundly alter the writer's own model of perception and description.

Many nineteenth-century treatments of the artist are concerned with his power to imitate the world. In *The House of the Seven Gables,* Holgrave's daguerreotype of Judge Pyncheon, for example, is credited with recording the real and unpleasant truth of the man's character, in contrast to the face Pyncheon wears to the world at large. And Drowne's wooden image, a figurehead carving in the Hawthorne story, achieves a likeness that defies comparison with the real thing. But at the same time that

[3]See Warner Berthoff, *The Ferment of Realism* (New York: Free Press, 1965), p. 8.

Hawthorne inflates the artist's capabilities in rivaling creation, possibly as a way of compensating for the actual neglect of the artist in American society, he also constructs tales that mirror his uneasiness about the artist's mimetic powers, especially when the lure of reproducing the real thing—through art or craft or science—results in the severing of the artist's ties with the world. That of course is the story of Aylmer in "The Birthmark," and Hawthorne's judgment of his perfectionist scheme of creating the ideal woman is clear: nature "permits us, indeed, to mar, but seldom to mend, and like a jealous patentee, on no account to make." A similar moral underlies Poe's "Oval Portrait," where the driven artist puts the finishing touches on the portrait of his bride, steps back and declares: "This is indeed *Life* itself!" The portrait may achieve life, but at the very moment the bride is dying.

Poe and Hawthorne are not in the tradition of realism—certainly not in the narrow sense of that word—but their preoccupation with the cognitive and spiritual dimensions of representation anticipates the strain of the American realist tradition that I am considering. E. H. Gombrich points out that "there are stories all over the world of images that had to be chained to prevent their moving of their own accord and of artists who had to refrain from putting the finishing touch to their paintings to prevent the images from coming to life."[4] The American contribution was to depict the mimetic process as a risk. Hawthorne's anxiety is lest the artist rival too closely nature's creation through his art, achieve a realism or perfection of creation that is too close to life itself; and behind this lies perhaps not only the danger of the artist usurping God's creative powers, but the Puritan heritage that gives priority to spiritual truth over the images and shadows of divine being. As Owen Warland learns, in "The Artist of the Beautiful," "When the artist rose high enough to achieve the beautiful, the symbol by which he made it perceptible to mortal sense became of little value in his eyes while his spirit possessed itself in the enjoyment of the reality."

If a concern with the artist's literal, mimetic powers of reproduction characterizes the first phase in the writer's anxiety, Hawthorne and James—sharing that concern—occupy opposing ends of the same spectrum. Hawthorne's uneasiness about the mimetic powers of the artist has become, by the time of James, an anxiety about the need for the artist to construct as perfect a perceptible symbol as possible. James has no doubts about the artist's vocation, and in fact his effort is precisely to rival the technology of the painter (not the photographer), who could provide an illusion of life: "The air of reality," he declares in "The Art of Fiction," "seems to me to be the supreme virtue of a novel. . . . It is here in very

[4]E. H. Gombrich, *Art and Illusion* (Princeton: Princeton Univ. Press, 1961), p. 111.

truth that he competes with life; it is here that he competes with his brother the painter in *his* attempt to render the look of things, the look that conveys their meaning, to catch the colour, the relief, the expression, the surface, the substance of the human spectacle."[5] But for James, still part of this first phase I am describing, the artist's successful creation of the illusion of the real thing is different from any literal imitation of the real thing itself. This is the subject of his story, "The Real Thing." Fallen aristocrats do not necessarily make as effective models of aristocrats—for the artist—as do vulgar but skillful models. His view is consistent with Thoreau's earlier denigration of the merely literal, scientific description, in contrast to the "true description" growing out of the perception of the object, which is "itself a new fact, never to be daguerreotyped,"[6] and it is consistent likewise with Whitman's ideal of rivaling material creation, but not by uselessly "daguerreotyping the exact likeness."[7] The precise terms of James's stoy may even have been something of a commonplace in the argument over realism by 1893, when it was published. Thus in 1883, Charles Dudley Warner had put the case strongly for "typical" realism in terms very similar to James's. In a piece in the *Atlantic Monthly* called "Modern Fiction," he wrote: "One of the worst characteristics of modern fiction is its so-called truth to nature. . . . A photograph of a natural object is not art; nor is the plaster cast of a man's face, nor is the bare setting on the stage of an actual occurrence. Art requires the idealization of nature. The amateur, though she may be a lady, who attempts to represent upon the stage the lady of the drawing-room, usually fails to convey to the spectators the impression of a lady. She lacks the art by which the trained actress, who may not be a lady, succeeds."[8]

Although Warner and Thoreau and Whitman used the photographic image as a metaphor for literal, unartistic imitation, in fact the notion of art as an idealization of nature was current among many photographers of the late nineteenth century as well. Art, Henry P. Robinson declared in 1892, "is not so much a matter of fact as of impression."[9] And Peter H. Emerson, another photographer, writing in 1880, distinguished between the scientific and artistic views. In portraying his subject, the artistic

[5]Henry James, "The Art of Fiction," in *The Future of the Novel,* ed. Leon Edel (New York: Vintage, 1965), p. 14.

[6]Henry David Thoreau, *Writings* (Boston: Walden ed., 1906), XX, p. 118. Quoted in Charles Feidelson, Jr., *Symbolism and American Literature* (Chicago: Univ. of Chicago Press, 1953), pp. 138–39.

[7]Walt Whitman, "Democratic Vistas," *Complete Prose Works* (New York: 1914), p. 244. Quoted in Feidelson, p. 3.

[8]Charles Dudley Warner, "Modern Fiction," *Atlantic Monthly,* 51 (April 1883), 464.

[9]Henry P. Robinson, "Paradoxes of Art, Science, and Photography," *Wilson's Photographic Magazine,* 29 (1892), 242. Rpt. Nathan Lyons, ed., *Photographers on Photography* (Englewood Cliffs: Prentice Hall, 1966), p. 82.

photographer must be wary of the scientific view, which "looks . . . so closely, so microscopically, at the butterfly's wings that it never sees the poetry of the life of the butterfly itself."[10]

Against the main current of idealized or typical realism in the nineteenth century, which had assimilated the photograph to its methods, there was a minor current of those who celebrated precisely its factual ability to record the skin's wrinkles, the marks on the drum head. Oliver Wendell Holmes was one of these, though his effort to improve the audience for photography in America ran against the popular taste, and his views would not prevail among serious artists until the twentieth century. Thus Holmes, writing in 1859 on the choosing of stereographs, advises investors to beware of group portraits, which counterfeit the upper class: "mostly they are detestable—vulgar repetitions of vulgar models, shamming grace, gentility, and emotion, by the aid of costumes, attitudes, expressions and accessories worthy only of a Thespian society of candle-snuffers. In buying brides under veils, and such figures, look at the lady's *hands*. You will very probably find the countess is a maid-of-all-work."[11]This is evidence, by the way, that lower-class models did not always succeed as well as they do in the Henry James story—at least in photography, where the camera's objectivity exceeds the painter's; but even Holmes is apparently not interested in the maid-of-all-work as a subject in itself. If photography was valued as a *truthful* medium, truth, for someone like Holmes, was restricted to certain subjects. It was only with the enterprise of Jacob Riis in the late nineteenth century that the poor became a subject worth attending to and that the photograph's capacity for detail would uncover a phase of reality previously unseen.

As long as photography was content to mimic the conventions of painting, its impact on literary realism was relatively small and its literal mimetic capacity ran against the symbolic tendencies of writers like Thoreau and Whitman. But by the end of the nineteenth century the journals are filled with debate on the threat to art—idealizing art is generally meant—posed by the more literal, documentary uses of the camera. (These new uses fostered an opening up of subject matter for the realist, mainly by attention to low life subject, and initiated a belief that would dominate the thirties—that the lower the life depicted, the more real the art.)

This change in the uses of the camera and its consequent impact on literature is dramatically embodied in the career of Alfred Stieglitz, who himself underwent a crucial change from pictorialism, the idealization of

[10]Peter H. Emerson, "Science and Art," *Naturalistic Photography*, 3d. ed. rev. (New York: Scovill and Adams, 1899), pp. 67–79. Rpt. Lyons, ed., p. 65.

[11]Oliver Wendell Holmes, "The Stereoscope and the Stereograph," *Atlantic Monthly*, 3 (June 1859), 747.

the photographic image along the lines of painting, to a more literal recording of reality. Through his own photographs and through the widely influential periodical, *Camera Work,* which Stieglitz edited from 1903 to 1917, the evolution of the photographic image from soft to hard, from idealized to realistic, would be broadcast. Stieglitz's impact on the avant-garde writers and artists of the early twentieth century, which has been documented by Dickran Tashjian and Bram Dijkstra,[12] was profound, and his example coincided with a fundamental reversal in the nature of realism that initiates what I have called the second phase of the realist's anxiety. In a world that seemed increasingly flooded with the identical products of mass industry and that was increasingly removed from direct experience through the hollow rhetoric of political discourse and advertising, Stieglitz restored the sense of real things. As Lewis Mumford put it in *Technics and Civilization,* after praising Stieglitz, "The mission of the photograph is to clarify the object. This objectification, this clarification, are important developments in the mind itself: it is perhaps the prime psychological fact that emerges with our rational assimilation of the machine. To see as they are, as if for the first time, a boatload of immigrants, a tree in Madison Square Park, a woman's breast, a cloud lowering over a black mountain—that requires patience and understanding. . . . Restoring to the eye, otherwise so preoccupied with the abstractions of print, the stimulus of things roundly seen as things, shapes, colors, textures, demanding for its enjoyment a previous experience of light and shade, this machine process in itself counteracts some of the worst defects of our mechanical environment."[13]

This restoration of a kind of Adamic realism took the form for writers of the twenties, not only of Hemingway's prose style, with its effort to capture "the real thing," but of Marianne Moore's abhorrence of rhetoric, her prescription that the poet include "real toads in imaginary gardens," and her cognate sense that "For anyone with 'a passion for actuality' there are times when the camera seems preferable to any other medium."[14] One might compare this view with the quite different one expressed in the 1890s by Howells' imaginary critic, who scorns the novelist's effort to describe an actual grasshopper, saying that if it's not commonplace, "you'll have to admit that it's photographic."[15] If a grasshopper was

[12]See Dickran Tashjian, *Skyscraper Primitives: Dada and the American Avant-garde, 1910–1925* (Middletown: Wesleyan Univ. Press, 1975); and Bram Dijkstra, *The Hieroglyphics of a New Speech: Cubism, Stieglitz, and the Early Poetry of William Carlos Williams* (Princeton: Princeton Univ. Press, 1969).

[13]Lewis Mumford, *Technics and Civilization* (1934; rpt. New York: Harcourt, Brace & World, 1963), pp. 339–40.

[14]Marianne Moore, *A Marianne Moore Reader* (New York: Viking, 1965), p. 187.

[15]William Dean Howells, *Criticism and Fiction* (New York: Harper and Brothers, 1892).

generally not welcome in the literature of 1890, a toad was by 1920. To the writer of the twenties, the camera would seem a model because its images had come to seem "more real" than reality itself.

Whatever the changes between the twenties and the thirties—and they are many—what remains constant among writers of the second phase, as I am describing it, is the centrality of the camera as a model for realism: the camera cannot lie. Bazin attributes this belief to the particular psychology of the photograph and the fact of its machine fabrication: "For the first time an image of the world is formed automatically, without the creative intervention of man. . . . In spite of any objections our critical spirit may offer, we are forced to accept as real the existence of the object reproduced."[16] The photograph permeated the documentary sensibility of the thirties, and while one result of the popular realistic arts of the time may have been, as Robert Warshow has complained, to move us "so close to reality as to destroy the detachment of art,"[17] the camera at its best could make the familiar unfamiliar. By its intense focus and selection, it could make us see things as if for the first time. It was precisely this sense of the camera's possibilities that had so great an impact on James Agee, who has come to seem a typical embodiment of thirties sensibility. Agee's *Let Us Now Praise Famous Men* evinces in its most acute form the anxiety of the realist who would emulate the camera's supposed objectivity, but Agee's is a creative anxiety, one that achieves a new level of reflexive consciousness and a consequent new level of realism.

Agee begins his meditations on the art of representation—which is the subject of *Famous Men*—by taking the extreme anti-art position: "If I could do it, I'd do no writing at all here. It would be photographs; the rest would be fragments of cloth, bits of cotton, lumps of earth, records of speech, pieces of wood and iron, phials of odors, plates of food and of excrement."[18] In short, the ideal is a sensational documentary, devoid of the intervention and distancing of art. But Agee cannot withhold himself from the effort of representation, and again and again he returns to the image of the camera as the ideal of objectivity, of detached presentation, for, as he puts it, the camera, "like the phonograph record and like scientific instruments and unlike any other leverage of art, [is] incapable of recording anything but absolute, dry truth" (211). (We should note that

Rpt. Eugene Current-Garcia and Walton R. Patrick, eds., *Realism and Romanticism in Fiction* (Chicago: Scott, Foresman and Company, 1962), p. 44.

[16] André Bazin, "The Ontology of the Photographic Image," *What Is Cinema* (Berkeley: Univ. of California Press, 1947), p. 13.

[17] Robert Warshow, "The Legacy of the 30's," *The Immediate Experience* (New York: Atheneum, 1970), p. 38.

[18] James Agee and Walker Evans, *Let Us Now Praise Famous Men* (1941; rpt. New York: Ballantine, 1966), p. 12. Subsequent references incorporated into text.

Agee is talking about a particular kind of photographer—epitomized by his collaborator Walker Evans—who handled the camera "cleanly and literally in its own terms" [211]. He was not talking about Margaret Bourke-White, for example, whom he later permits to hang herself in an interview reprinted at the end of the book.)

Agee's reverence for the camera's objectivity is part of a larger reverence for a way of knowing the world that encompasses the ideal of a humanistic science. The particular realism that results from this reverence for science is a way of seeing, Agee says, "of which the still and moving cameras are the strongest instruments and symbols. It would be an art and a way of seeing existence based, let us say, on an intersection of astronomical physics, geology, biology, and (including psychology) anthropology, known and spoken of not in scientific but in human terms" (211). We can only say that Agee fails to achieve his goal: his literal descriptions—to take him at his most objective—may emulate the camera, but their effect is to create a world of sounds and words and associations that is simply incomparable—as any verbal text must be—to the still photograph. And certainly the whole of *Famous Men* is tonally far different—more baroque, more self-conscious, more subjective—from the stated ideal of objective, scientific knowledge, far different too from the direct simplicity of Walker Evans, if we can make any comparison at all. So Agee is doomed to fail: with our present certainty of Heisenberg's uncertainty principle, we can regard Agee's belief in the objectivity of the recording instrument as naively innocent. But paradoxically Agee succeeds precisely by virtue of the effort that is doomed to fail; his heightened sense of realism springs from the intense self-scrutiny and reflexive consciousness that he brings to the act of observation rather than from the stated ideal of objectivity. He succeeds, we might say, by virtue of his inadvertent adherence to the Heisenberg principle, that is, by taking account of the methods of observation and limitations to objectivity. Not Evans, but Whitman is his true mentor, whose goal of creating a path between reality and the soul Agee achieved better than the poet himself perhaps.

But if Agee's record of the Alabama tenant farmers testifies to the fertility of the realist's anxiety to emulate the camera, it signals a turning point of another kind. Agee adds a subtle twist to the anxiety of realism. For all his conscious emulation of the camera's objectivity, he is also extremely wary of it, in a way that other documentarists—from Jacob Riis to the FSA group under Roy Stryker—were not. For it is Agee who observes acutely that Walker Evans' depiction of the victims of poverty could itself be a kind of victimization, with the photographer "stooping beneath cloak and cloud of wicked cloth, and twisting buttons; a witchcraft preparing, colder than keenest ice, and incalculably cruel"

(332); and Evans' subjects realize it too (or so Agee imagines), understand "the meaning of a camera, a weapon, a stealer of images and souls, an evil eye" (329). And insofar as Agee shared the goals of "objective" documentation, he shared too Evans' guilt at the act of aggression implicit in the artist's effort to establish mimetic proximity.

From the beginning, the camera's ability to record *everything* made people uneasy. A writer in the 1840s, for example, while excited at the prospect of picturing trees budding and flowers going to seed, sounds more than a trifle discomfited at some other uses of the camera: "A man cannot make a proposal or a lady decline one . . . or a reverend bishop commit an indiscretion, but straightway, an officious daguerreotype will proclaim the whole affair to the world."[19] And an acute observer in the 1890s projected fears that would become commonplace only later when he declared, looking into the future, that with the taste for actuality growing, we shall soon be "besieged at every turn by the litterateur and artist eager for material and armed with camera and photograph. Hidden automatic machines will catch the unwary in every word and act: walls will have ears and ceilings eyes. . . ."[20] What is significant about Agee's anxiety is that he moves it from the potential victim of the camera's omniscience to the perpetrator, the recorder himself, and needless to say he is at least as guilty of intrusion into his subjects' lives as a writer as Evans is as a photographer.

Agee's anxiety is of another sort as well. In his effort to achieve the perfect realism, the record of a true sight of the world, the artist must not be guilty of using the traditional tools of literary emphasis, of exaggerating for effect. "You abjure all metaphor, symbol, selection and above all, of course, all temptation to invent, as obstructive, false, artistic" (212). Here too there is some continuity with certain writers of the twenties (Williams and Moore and the studious avoidance of metaphor), but Agee's anxiety is very different from that of James, for example, whose notebooks are filled with the moment by moment strain of inventing the successful illusion, and who feared just the opposite: not being "artistic" enough.

One possible reason behind the rediscovery of Agee in the late sixties and seventies is that his anxiety is still to some extent with us, coloring the social scientist's sense—if not that of the creator of mass culture—of just how objectively he may be observing, as opposed to making literature. That this remains a significant theme in recent literature may be seen in Cortazar's story, "Blow-Up." In that story, Cortazar presents an image of

[19] *The Living Age*, 9 (1846), 551. Quoted in Robert Taft, *Photography and the American Scene* (1938; rpt. New York: Dover, 1964), p. 68.

[20] Hiram M. Stanley, "The Passion for Realism, and What Is to Come of It," *The Dial*, 16 April 1893, p. 239.

the confrontation of the photographer with his subjects, or more exactly, an image of the photographer's confrontation with the image of that confrontation—the candid shot he has taken of a wary and uncandid trio. In commencing his narrative, the photographer Michel wishes, interestingly, that his typewriter could interpret the results of his camera, for "one machine may know more about another machine than I, you, she—the blond [whom he's photographed]—and the clouds." But that cannot be: Michel is compelled to interpret the reproduction and to search for its hidden meaning—assuming it has one. The subject of the story is not so much what Michel eventually decides is the "truth" behind the image, as the writer's troubled imagination in the act of interpreting the world, his expressed fear, as Michel puts it, of being "guilty of making literature, of indulging in fabricated unrealities."[21] For Cortazar too, apparently, the camera, because it is a machine, cannot lie, while the act of writing interposes a subjectivity, a "literary" way of seeing that is distrusted in the face of the machine's superior objectivity.

Looking back to the nineteenth century, one can see the modern epistemological anxiety of Agee and Cortazar as a complete reversal of James's and Warner's avoidance of the supposed literalism of the camera. But the contemporary writer—to move now into the third phase in the progression I am describing—goes beyond this modern epistemological anxiety. What distinguishes the contemporary realist from his predecessors is that while still being concerned with knowing and describing the world and the camera's way of imaging the world, he has ceased to worry about the literal correspondence of his verbal description with an apparently more literal visual image, has lost the anxiety of realism that has been so characteristic of the mode. His sensibility reflects a new attitude toward technologies of reproduction—a new skepticism. I cannot document precisely when this turning toward the avant-garde contemporary attitude took place, but one can see it reflexively depicted in a recent story which in fact is about the anxiety of realism. Keith Fort's "The Coal Shoveller," originally published in 1969, depicts a budding writer trying to depict, as faithfully as possible, a "realistic" subject—a figure from the traditional cast of realism, the lower class, a romantically "real" laborer—the coal shoveller. What it also depicts—and this recalls the mode of Agee's realism—is the artist's struggle self-consciously to come to terms with the nature of verbal expression. But what marks the new sensibility is the attitude the author takes toward the realistic enterprise, which is not one of anxiety, but rather a mocking of anxiety. Thus, after one false attempt at describing the black man shovelling snow outside his window, he stops short:

[21]Julio Cortazar, "Blow-Up," in *End of the Game and Other Stories* (New York: Random House, 1967). Rpt. Philip Stevick, ed., *Anti-Story* (New York: Macmillan, 1971), p. 163, p. 171.

> I could write a hundred essays on how objective I am going to be, but the connotations of 'white,' 'snow,' and 'black' would still be there. . . . Can someone like Robbe-Grillet actually believe that he is reshaping the mind of man as he writes? I don't have the arrogance to deny reality in the name of an idea of reality.
>
> Nor am I convinced that fiction *should* try to approximate painting (or the films). Prose has been headed in this direction for some time, but like most movements in art this one was begun on ill-defined premises and succeeding writers merely spun out a potential. We have come to a point where the weaknesses inherent in the original assumptions prevent us from going further. To ask words to make fiction into photographic realism is to demand a performance which they are totally incapable of giving.[22]

After several more false attempts, the would-be author adopts self-satire as the only viable mode. He imagines sending himself out to a local bar, there to encounter the actual shoveller, with whom he can naturally establish no rapport whatsoever. The story ends in the fine dilemma of the writer, sitting at his desk as the light outside grows dim, unable to turn on his room light because it would prevent him from seeing his subject, and consequently unable to see the paper he is writing on—a wonderful image of the contemporary realist's paralysis. But the comic tone of the piece points the way to the new freedom of the writer without anxiety who will perhaps turn on the television set, watch a concerned interview with the same coal shoveller, and, happily whistling in the dark, tell us what he sees. Perhaps there will be a knock on the door. It is the coal shoveller, with a copy of his photograph in the latest *Newsweek*.

The contemporary writer who best exemplifies the sensibility of the third phase in the evolution is Donald Barthelme. With his omnivorous appetite for the images and idioms of mass culture, Barthelme is the representative voice of our time, as Oldenburg or maybe Red Grooms is the representative sculptor, and in all of them a similar sensibility prevails. Barthelme's works especially are filled with the data of our mass-produced environments; the brand-names are named, and so on occasion are real people, as in "Robert Kennedy Saved from Drowning." Philip Stevick defines this quality exactly when he calls Barthelme's work a literature of "mock fact."[23]

But what I would like to call attention to specifically in Barthelme—

[22]Keith Fort, "The Coal Shoveller," *The Sewanee Review*, 77, No. 4 (Autumn 1969). Rpt. Stevick, ed., pp. 41–42.

[23]Philip Stevick, "Lies, Fictions, and Mock-Facts," *Western Humanities Review*, 30, No. 1 (Winter 1976), 2–12. See also the shrewd definition of Barthelme's "anironic attitude" in Alan Wilde, "Barthelme Unfair to Kierkegaard: Some Thoughts on Modern and Postmodern Irony," *Boundary 2*, 5 (Fall 1976), 45–70. Martin Greene anticipates the contemporary sensibility in his discussion of Nabokov and Salinger in "American Rococo," a chapter in *Re-Appraisals* (New York: Norton, 1965).

and it's a quality he shares with several other contemporary writers—is the freedom with which not only real things, but images of real things are incorporated into the fiction. As the photograph and the moving picture have become more and more a part of our daily lives, they have ceased to function for us as mirrors of the real world; instead, they have become a part of the landscape that we critically scrutinize. Their relationship to "the real world" has become problematical, as two recent Barthelme fictions illustrate.

One of the pieces in Barthelme's miscellaneous collection, *Guilty Pleasures,* is entitled "The Photographs." At the head of the first page are two images that seem to be engravings (not photographs) of the same object—one a top view and one a side view. The object appears dish-like, it has a handle, and is possibly made of stone, or feathers or papier-mâché. What is it? Barthelme tells us right away:

> The attached photographs of the human soul (Figs. 1 and 2), taken by Pioneer 10, the first spacecraft to navigate the outer solar system, were made on December 4, 1973, as the craft was leaving the magnetic fields of Jupiter. The "photographs" (actually coded radio signals from the device's nine-foot dish antenna beamed back to earth) were, of course, incidental to the photographing of Jupiter itself, one of the mission's chief aims. They were made by Dr. Reginald Hobson, F.R.S., of Britain's Cavendish Laboratory, using Kodak spectroscopic plates type IIIa-J baked for five hours at 65° C. under dry N_2 before exposure. Dr. Hobson very shortly afterward brought the resulting images to his friend and colleague Dr. Winston Watnick-Mealie, F.R.S.[24]

The rest of the piece, following this parody of factual data, consists of a dialogue between the two scientists mentioned, who debate the meaning of the photographs, deciding finally that they are pictures of "the human soul on its way to Heaven," though they look more like corroded frying pans. While discussing whether or not such ugly pictures of the soul ought to be revealed (the responsibility of the scientist and so forth), one of the pair reveals the rather more personal and immediate stain on his own soul—the fact that he has been having an affair with the other's wife. The two scientists conclude, with only slightly altered civility, that it would be better to burn the photos of the soul than to publish them. "Pity they weren't prettier."

This ending may leave us in some doubt about the authenticity—if that is the word—of the initial "photos" and accompanying data (how did they survive the burning?) but that little puzzle is not really the point. Rather, Barthelme is beguiling us with the notion of an impossibility, playing with the language of science and with the distance between that language and emotions, and dramatizing the space between a photo-

[24]Donald Barthelme, *Guilty Pleasures* (New York: Farrar, Straus and Giroux, 1974), p. 153.

graphic image and an abstract concept. For you obviously cannot image the human soul, although Hawthorne came close in the raised flesh of Dimmesdale.

Barthelme's story mocks our universal faith in photographic truth and in the camera's ability to record the secrets of the universe, from outer galaxies to inner organs; and it is paralleled by another piece, called "A Film," which includes a report on the filming of the moon rocks at the Smithsonian. The utterly disappointing, dumb inertia of the rocks is transformed, in the eyes of Barthelme's filmmakers, into "the greatest thing we had ever seen in our entire lives! . . . Peering into the moon rocks, you could see the future and the past in color, and you could change them in any way you wished. The moon rocks gave off a slight hum, which cleaned your teeth, and a brilliant glow, which absolved you from sin. . . . The dynamite out-of-sight very heavy and together moon rocks turned us on, to the highest degree. There was blood on our eyes, when we had finished them."[25] In this pop dithyramb Barthelme is clearly having fun with the contemporary idiom, but more to the point here is his playing with our notion of the film medium as a vehicle for mass-fantasies, transforming the humdrum moon rocks into dazzling virtuosos through the fact of their star status.

Another contemporary example of the problematical relationship between mass-produced image and "the real thing" is a work by Les Levine which appeared in *TriQuarterly* 32 (Winter 1975). The first two pages of the Levine piece (I am not sure whether to call it a fiction or an essay) present the reader with a large outline of an actual photograph of Nixon in China. Nixon is eating with Chou En-lai and others at a round table, and key numbers are drawn over certain points in the image. Succeeding pages present close-ups of those numbered points, with corresponding paragraphs—ranging from a few lines to a page in length—probing the possible meanings of the image: the political figures are themselves of course images, representatives; their postures and demeanors are signs with political overtures; and the central mystery and gesture of the image—Nixon picking up a cashew nut with a pair of chopsticks—is made a leitmotif, recurring throughout the analysis as a symbol of international relations. Levine's tone is chatty and personal, curious and casual, including a number of comments on the formal, artistic nature of the photography and on the relationship of form to

[25]Donald Barthelme, *Sadness* (1972; rpt. New York: Bantam, 1974), pp. 77–78. An early anticipation of Barthelme's parodic treatment of film and the mass media occurs in Elmer Rice's *A Voyage to Purilia* (New York: Cosmopolitan Book Corporation, 1930), which satirizes many of the conventions of acting and camera movement in early movies, as well as the stereotyped roles and situations that constituted the fantasy of American life for the mass audience of the 1920s.

meaning. The whole is a parody of allegorical interpretation (and possibly too of popular culture studies) that, despite its excesses, does not at all exhaust the possibilities for interpretation. Rather, we begin to lose ourselves in the world of the image as meaning after meaning is opened up, and we may end by sensing the final opacity of the photograph.[26]

In thus making a plausible fiction out of his reading of an actual photographic image, Levine returns to one of the original sources of fascination with the mechanically reproduced image. As one writer observed of the photograph around 1850, it was "A perfect transcript of the thing itself."[27] And the painter Samuel French Morse, writing in 1839 in anticipation of Levine's method, reflected on his use of a microscope to inspect a daguerreotype: "The effect of the lens upon the picture was in a great degree like that of the telescope in nature."[28] And yet, obviously, between the mid-nineteenth century and our own time a radical reversal has taken place: where the nineteenth-century viewer would see the image as the mirror of the real thing, a window onto the world outside, the contemporary writer uses the photograph as a base for conjectural analysis, an exercise field for the imagination and for speculation on the relationship between image and real thing.

Knowing furthermore that the verbal medium cannot achieve "photographic realism" the writer has freedom to play with concepts of correspondence in a way that would be alien to Agee's generation. It is a freedom that has marked the contemporary representational arts generally. Thus, during the last ten years, the painters often grouped together as super-realists have likewise taken the photographic image as subject matter, exploring its iconic significance in our culture and also raising questions about perception and the technology of reproduction. Photographic realism—in the sense of an unquestioned correspondence between image and subject—is, for both the writer and the painter today, a naive fiction. In a sense the artist arrived at this suspicion later than the social scientist. For thirty years students of anthropology and of visual communication have been aware of the subjectivity of the photographic medium and have studied the stylized vocabulary and covert rhetoric of ethnographic documentaries. As a well-known text in the field puts it, "one can regard the photograph as we use it as an arbitrary linguistic

[26] A more extreme contemporary example than Levine's of reading (or rather *not* reading) an image is provided by Duane Michals in his photography of a barroom, which is framed, top and bottom, by a handwritten text—really a kind of anti-caption—noting some elements in the photographer's experience "on location" that do not show up in the image. It is called, *There Are Things Here That Are Not Seen in This Photograph* (1977).

[27] *The Diary of Philip Hone, 1828–1851,* ed. Bayard Tuckerman (New York: Dodd, Mead and Company, 1889), I, p. 391. Quoted in Richard Rudisill, *Mirror Image: The Influence of the Daguerreotype in American Society* (Albuquerque: Univ. of New Mexico Press, 1971), p. 53.

[28] S. F. Morse, letter to *New York Observer,* March 1839. Quoted in Taft, p. 12.

convention not shared by all peoples."[29] And the visual studies journal *Afterimage* recently reprinted an article written around the time that Cortazar and Fort were writing their fictions, that exactly shares these writers' concerns: "the photographer is *always* a part of the context of the events he is photographing; he can never photograph a human behavior without being a part of it."[30]

That point is implicitly underscored in the recent work of the photographer Duane Michals, many of whose image-series incorporate himself as photographer. (See, for example, *The Illusion of the Photographer,* which begins with several photographs of a street scene with a human figure moving across it, and then moves back to show a view of the same scene through a proscenium arch; the series ends by showing us the photographer himself taking the image through the mock-stage and finally turning to look at the viewer in the last frame.) Michals' work has come increasingly to blend visual image and literary text (in the artist's handwriting, surrounding the print and commenting upon it) and approaches—from the photographer's viewpoint—the same terrain of mixed media occupied by writers like Barthelme and conceptual artists like Levine. The premise is the same for all three: the camera is not an instrument for recording "objective truth." One relevant example of Michals' confrontation with the metaphysics of photography is the recent non-photograph consisting solely of the following caption:

> A Failed *Attempt to Photograph Reality*—How foolish of me to believe that it would be that easy. I had confused the appearances of trees and people and automibles [sic] with reality itself and believed that a photograph of those transient appearances to be a record of it.[sic] I am a reflection photographing other reflections within a reflection. It is a melancholy truth that I must always fail. To photograph the ultimate reality is to photograph nothing.

There *is* something melancholy about the "truth" here affirmed, but the tone of the assertion also mocks the sentiment and the result is, one might say, a lighthearted anxiety. Michals' affinities with a writer like Barthelme are evident.

In the contemporary climate of mimetic uncertainty the artist has felt a liberation from the constraints of accuracy and realism and genre that is different from the more traditional liberation of fantasy. Fiction and non-fiction, made art and ready-made art, life and art, have become blurred categories. Writers of documentaries—Mailer, Tom Wolfe—frankly admit the use of fictional techniques to add a more vivid feel for the texture of

[29]Marshall H. Segall et al., *The Influence of Culture on Visual Perception* (Indianapolis/New York: Bobbs-Merrill, 1966), p. 33.

[30]Paul Byers, "Cameras Don't Take Pictures," *Columbia University Forum* (Winter 1966); rpt. *Afterimage,* 4, No. 10 (April 1977), 9.

reality; and the writer of fiction—Doctorow's *Ragtime* and Coover's *Public Burning* come to mind—freely imports actual persons into his fictional frame, without pretense of accuracy. (All of this is of course different from the traditional realistic novelist's claim of utter veracity in his tale; it is different too from Dos Passos, who juxtaposed biographies of real persons with fictional narratives, but who was scrupulously careful to isolate the various ontologically different persons in separate formal chambers.) The contemporary writer's freedom to mix the real and the fictional, and his insouciant, parodic tone are acutely symptomatic of our present sense that we live in a world of images, that the mass produced items of consumption have made our physical environment a series of duplications, and that mass images of contemporary culture—from the Warhol soupcan (now emblazoned on plastic pillows) to the "docu-drama"—have made questions about "the real thing" at once obsessive and irrelevant.

First the realist worried about his competition with the painter; then with the photographer. And the competition of the realist with journalism and the mass media was foreseen as early as 1894, when a critic observed that "if *realness* be the final test," a novelist like Howells, for example, cannot compete with the "really real heroes of the newspapers."[31] But the contemporary realist, alive to the mass idioms of the American environment, does not need to compete with photography or the mass media anymore, because "realness" is no longer assumed to be inherent in the various mimetic modes. In this climate of confusion between fact and fiction, between information and entertainment, the realist has largely lost the anxiety that characterized his predecessors, has lost the anxiety of competing with supposedly "truer" technologies of mechanical reproduction. But he does, through his art, still picture the world in a way that challenges us to see it in all its confusions. If the irresistible tendency of technology—whether of mass production or of mass communication—is to broaden the social base of what is familiar, the function of the realist is to make us see the familiar with new eyes, to make it strange.

[31]William R. Thayer, "The New Story-Tellers and the Dawn of Realism," *Forum*, 18 (December 1894). Rpt. Current-Garcia and Patrick, eds., p. 159.

Notes on the Technological Imagination

MICHEL BENAMOU

At the time of his death in the spring of 1978, Michel Benamou was developing what he called a "subjective square" of discourse about technology, a typology of attitudes that he also described as "the four humors of discourse about technology." Unfortunately, he had only begun this research. I have selected the following excerpts from Benamou's working papers. They present his map of the "four humors," sketch out two of its basic positions, and suggest the comparison he intended to make between discourse about technology and literature that reflects a technological "mind-set."

I. *Introduction*

At the beginning of *Calligrammes,* Apollinaire hailed, prophetically, the informational society. Addressing all the communications media—rails, transatlantic cables, radio waves—he intuited that they would be like the meshes of a net: ambiguous *liens* of both intimacy and bondage. Relating these new technologies to the body, he greeted their inroads as heralding a new sensuality:

> J'écris seulement pour vous exalter
> O sens ô sens chéris
> . . .
> Ennemis de tout ce que j'aime encore.[1]

Done: man has struck his Faustian bargain with technology. Tech-

[1]Guillaume Apollinaire, *Calligrammes* (1925; rpt. Paris: Gallimard, 1966), pp. 23–24.

nological man—protean, ubiquitous, communicating with the universe—has attained the new sensibility that Robert Jay Lifton describes as "omni-attention, the sense of contemporary man as having the possibility of 'receiving' and 'taking in' everything."[2] But a "massive disillusionment" also gnaws at this protean man, for he is worried by the destructive powers of technology. If we can call John Cage his prototype in the United States, then we can also observe that between 1912, the approximate date of the poem I have quoted, and 1977, the publication date of Cage's recent dialogues with Daniel Charles (*Pour les oiseaux*), the technological imagination has undergone a profound transformation.

By a transformation in the technological imagination I mean to suggest two phenomena. The first is that recent discourse about technology—what I call technocriticism—has become increasingly circumspect in its attitude toward postindustrial society. The second is that the preferred forms and metaphors of twentieth-century literature have become increasingly technological. An understanding of the issues raised by the former should allow us to adopt a rigorous critical stance with respect to the latter; an understanding of the latter should allow us to view the former in a larger cultural context. In both cases, my focus will be on works by Americans, on works which have become part of the American discourse about technology, or on works about the United States.

II. *American Technocriticism: A Subjective Square*

During the past ten to fifteen years over a hundred volumes on technology have appeared in the United States. They range from Lewis Mumford's *The Pentagon of Power* and Jacques Ellul's *La Technique* (translated only in 1966) to the Club of Rome's pronouncements and Victor Ferkiss's *The Future of Technological Civilization*. Besides their pragmatic and programmatic intentions, an urgency, almost a passion, distinguishes these books from the discourse about science. This impassioned tone belongs to the subject: although scientific knowledge is often pressed into service to justify models of political or economic behavior, discourse about science tends to mask ideology with its claim to objectivity. Discourse about technology, on the contrary, can make no such claim. Under the threat of extinction or the hope of millenary post-industrialism, it is almost always visceral, humoral, polemical. It fills the four corners of a subjective square which I have borrowed from William Irwin Thompson's *At the Edge of History* and have modified to contain what I call the four humors of discourse about technology:[3]

[2]Robert Jay Lifton, *Boundaries: Psychological Man in Revolution* (New York: Vintage, 1967), p. 51.

[3]See William Irwin Thompson, *At the Edge of History: Speculations on the Transformation of*

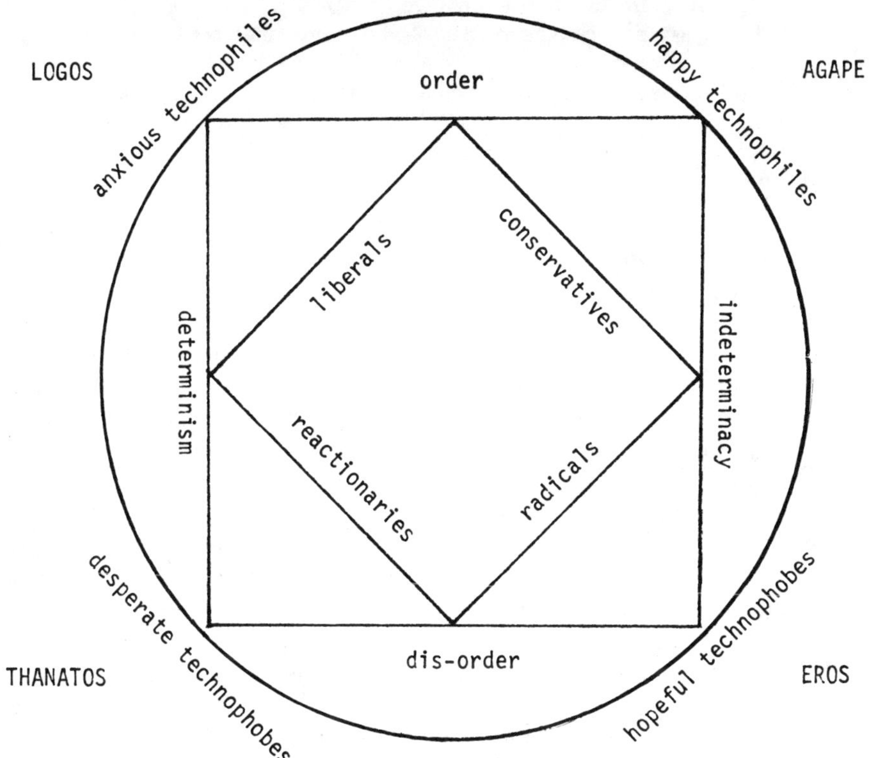

—*Happy technophiles* are politically conservative believers in *Agape,* such as Marshall McLuhan, Buckminster Fuller, and Paolo Soleri. Admirers of Teilhard de Chardin, they believe technology will bring about convergence and unification.

—The liberal worshippers of a rational *Logos* are *anxious technophiles,* who see contradictions in planetary culture but keep their faith in democratic controls over technology. Their prototype is the Lewis Mumford of *Technics and Civilization.*

—Disillusioned rationalists become *desperate technophobes,* who maintain that the worst technological disaster has already occurred in our Western minds. They include Jacques Ellul, who sees technique as autonomous, the Mumford of *The Pentagon of Power,* and the hopeless Marcuse of *One-Dimensional Man.* To them, planetary culture means the unavoidable destruction of basic human values, defined in terms of a stable relation to nature. Thus, they appear to be reactionary.

Culture (New York: Harper and Row, 1971), pp. 74–107, 166–70.

—The fourth corner is occupied by *hopeful technophobes*. They are the radical party of *Eros* against *Thanatos:* ecological anarchists like Paul Goodman, the surrealist Marcuse of *Essay on Liberation,* Ivan Illich, and Theodore Roszak. They are convinced that urban industrialism is but a failed experiment which will be replaced by human-scale technologies once the counter-cultural forces of the "Great Refusal" and the "New Sensibility" have prevailed.

In general, we can say that what Ihab Hassan has called a debate between technophiles and arcadians—a formulation that he too has seen as problematic—is complicated by internal contradictions: there are happy technophiles and anxious ones, desperate technophobes and hopeful ones. At times the same thinker shifts from one pole to another: for example, the Lewis Mumford of 1934 was an anxious technophile, but by 1970 he had become a desperate technophobe; Marcuse, on the other hand, desperate in his *One-Dimensional Man* of 1964, had become a more optimistic technophobe in his *Essay on Liberation* of 1969.

Two philosophical positions suffuse this discourse about technology: determinism and the closely related but distinct notion of the autonomy of technology. Their antidotes are indeterminacy and the political control of technology. Contributors to technocriticism, moreover, represent a variety of disciplines. They include sociologists (Bell), economists (Toffler, Heilbroner, Kahn, Schumacher), historians of culture (W. I. Thompson, Roszak, Mumford), communications specialists (McLuhan, Bagdikian), theologians (Illich, Cox), inventors (Fuller, Cage, Soleri), literary scholars (Hassan, Marx, Sypher, Goodman), and philosophers (Ellul, Hans Jonas). Whatever their differences, they all recognize the indisputable fact raised by McLuhan as early as 1948: the present generation is already living in a technological or postindustrial era.

A. *Happy Technophiles*

Norbert Wiener, the father of cybernetics, announced in 1950 a "Second Industrial Revolution" linked to electronic communication systems. Postindustrial society would differ from industrial society (based on capital) as much as the latter differed from agrarian society (based on land). The base of postindustrial society would be technological expertise, with know-how calling the shots. If the primary product of the industrial era was manufactured goods, that of the postindustrial epoch would be services. And, as Victor Fuchs and Daniel Bell have noted, in 1956 the United States became the first country in the world in which the number of employees in the tertiary industries exceeded those in production.[4]

[4]Daniel Bell, "Notes on the Post-Industrial Society," *The Public Interest,* No. 6 (1967), pp. 24–35. Bell, in turn, refers to Victor R. Fuchs, "The First Service Economy," *The Public Interest,* No. 2 (1966), pp. 7–17.

This moot opposition between goods and services, plus the myth that automation would raise the level of expertise of all workers (thus producing a classless society), plus discoveries in electronics (computers as conduits of global communication), nourished the visions of a swarm of technophiles promising inexhaustible honey. Primarily economists, such as Victor Fuchs and Herman Kahn, they are deterministic and optimistic in their belief that it is economic structure which transforms culture. If culture does not respond to the goad of technology, it is because of cultural inertia—a delay which the technocrats gloss over. Herman Kahn and his associates predict that the year 2,000 will see a new human type: super-rational at work, irrational in private life.[5] Primarily a refined consumer, this efficient sensualist is necessary to the progress of multinational corporations in a world at peace, a world so insatiable and so profitable that the standard of living will be raised tenfold for everyone, everywhere.

A similar determinism informs the thought of Marshall McLuhan: under the dominion of the media, civilization will become an electronic Eden. An extra-corporeal extension of the brain and the nervous system, television now "suits" the species in the same way that the book "suited" it as an extension of the eye. For McLuhan, an Aristotelianism determines the future: human nature is endowed with five senses, so television's equalizing of them must be a good thing. Apollinaire hailed nothing else: "ô sens chéris."

But it is difficult to decide, finally, whether we should read this prophet of the postindustrial era as an apologist for, or as a critic of, the electronic revolution. The ambiguity arises from McLuhan's agile side-stepping of the political question. While he joyfully anticipates the return of our aural-tactile nature after centuries of visual-typographical domination, while he preaches "human docility" and submission to electronic technology, his assent does not acknowledge the potential for a technocratic dictatorship.

Indeed, McLuhanism is actually an aesthetics rather than a social theory. Inspired by the observation that American youth is becoming increasingly resistant to the book, founded on Innis's and Carpenter's research in communication, fed on Shakespeare and T. S. Eliot, lacking any social or psychological theory but plugged into the organs of reception, McLuhanism envisions a global village of televiewers united by mass culture. But it has no response to a central question posed by other technocritics: "Who controls the media?"

Another apolitical and determinist technophile in the gallery of postindustrial prophets is Richard Buckminster Fuller, who captured the

[5]See Herman Kahn, Anthony J. Weiner, et. al., *The Year 2000: A Framework for Speculation on the Next Thirty-Three Years* (London and Toronto: Macmillan, 1967).

technological imagination with his metaphor of Spaceship Earth long before our two astronauts strolled on the moon. Hugh Kenner has devoted an entire book to "Bucky," in which he concludes that: "Metaphors, paradigms, these are our greatest needs, irradiating minds with heuristic images: points of departure, not solutions; encouragements to Dig Wholes."[6]

The problem is that these images are meant to function in the political arena as well as in the aesthetic context of design. Thus, the geodesic dome, according to Fuller, ought to do away with "localism and nationalism." For him, technology results not in an electronic Eden, as for McLuhan, but in a spatial Utopia: "The comprehensive introduction of automation everywhere around the earth will free man from being an automaton and will generate so fast a mastery and multiplication of energy wealth by humanity that we will be able to support all of humanity in ever greater physical and economic success anywhere around his little spaceship *Earth*."[7]

Fuller's wisdon, however, is part and parcel of his megalomania, for he knows that no specialized solution is adequate to the problems posed by technology. "The kind of technology that endangers," Fuller writes, "is that occasioned by the blinders of specialization where each of our various acts is executed without consideration of the others."[8] Post-industrial America must become the type for a global civilization that is technical, synergistic, liberated from *Dasein* through *design*, and in communication with millions of other inhabited planets. It is that, or annihilation. Utopia, or oblivion.

What hinders this technologic of history—the transformation of humanity by the media or by geodesic domes—is, of course, human perversity. Hence, we also have a body of futuristic literature favoring psychological modification: from B. F. Skinner's *Walden II* (1948), which champions the simple conditioning of social motivations, to Gerald Feinberg's *The Prometheus Project* (1969), which proposes access to a universal extraterrestrial consciousness as one of the objectives of biological engineering—a curious grafting of Alexis Carel onto Teilhard de Chardin.

Perhaps the attraction to astronautical metaphors that we find in such examples of the postindustrial imagination masks a profound disillusionment. The appeal to a new human type may imply a despairing

[6]Hugh Kenner, *Bucky: A Guided Tour of Buckminster Fuller* (New York: William Morrow & Co., 1973), p. 314.

[7]R. Buckminster Fuller, *Utopia or Oblivion: The Prospects for Humanity* (New York: Bantam, 1969), p. 362.

[8]R. Buckminster Fuller, "Technology and the Human Environment," in *The Futurists,* ed. Alvin Toffler (New York: Random House, 1972), p. 304.

view of postindustrial man's ability to come to grips with the disasters of pollution, overpopulation, technology-induced unemployment, and political totalitarianism. Neither Kahn and Co., McLuhan, Fuller, Skinner, nor Feinberg, we might say, propose a genuinely "critical" technocriticism. Their writing seems inspired by desire for order (planetary or even cosmic) rather than by social conscience. They represent the technological imagination of the postindustrial era at its apogee: euphoric, eudemonistic, and eugenistic.

B. *Anxious Technophiles*

Together, the crisis of capitalism dramatized by the limits of energy resources and a few books criticizing the economic growth of the West sufficed to restrain the technophilic burst of 1967. To the millenary maunderings of the Commission for the Year 2,000, for example, the Club of Rome opposed a first report, *Limits to Growth* (1972), and a second simulation produced on the computers at MIT, *Mankind at the Turning Point* (1974). The verdict was that the industrial era was going to end, not in an electronic revolution, but in the irreversible exhaustion of all the earth's resources! Such is the upshot of the global planners' enthusiasm for their computers: in less than ten years, postindustrial plenty gives way to widespread famine. There is an internal logic to this reversal: blind faith in machines—the science of systems included—is a reductionism destined to reduce itself. Moral: don't leave technology to the technologists.

But American technocriticism did not need the lesson of the oil crisis in order to react to the wave of optimism of 1967. Readers of history, particularly, had already been putting the postindustrial revolution in perspective. As early as 1959, in *The Future as History,* Robert L. Heilbroner had denounced the technophile illusion, prefigured in Adam Smith's *The Wealth of Nations,* of a determinism applied to predictions of the future. Heilbroner situated the embarrassment of riches in the American economy on a slope leading to the progressive erosion of individual autonomy by an encroaching industrial-military-political complex. In *The Human Prospect,* which appeared in 1975, Heilbroner attacked such Western values as capitalist profit and technical expertise for having failed to satisfy mankind. He opted instead for an ambiguous monastic order that would be both religious and military, for an economy and a technology based on the Chinese model, for an asceticism dictated by the impoverishment of the Earth. As Heilbroner put it: "In our discovery of 'primitive' cultures, living out their timeless histories, we may have found the single most important object lesson for future man."[9] Because Heilbroner is a non-determinist, he concedes a chance

[9]Robert L. Heilbroner, *(An Inquiry into) The Human Prospect* (New York: W.W. Norton,

for survival to Atlas—an Atlas who may usurp Prometheus's special position in the technological imagination.

Technological Man (1969) by Victor Ferkiss constitutes another ripost to the technophile Utopia. Reviewing such millenarian prophets and prophecies as B. F. Skinner and his conditioned Utopia, C. Landers and his astronaut-machine, Teilhard de Chardin and his noosphere, Arthur Clarke and his Cyborg, and Harvey Cox and technopolitan man—all consequences of the communications revolution—Ferkiss carefully distinguishes the postindustrial reality from the postindustrial myth. At least his *pataphysique* (science of imaginary solutions) is more poetic than the sensualist predictions of Herman Kahn, whose new human type Ferkiss denounces as a

> neoprimitive man trapped in a technological environment. . . . Rather than develop a new culture suitable for technological man, man may withdraw culturally before the existential revolution and end up living in cultural poverty and bondage, relieved by outbursts of frustration in much the same way as certain primitive groups (some American Indian tribes, for instance) who have been overcome by industrial civilization and deprived of a functioning culture.[10]

But are we in the midst of an existential revolution in keeping with technological transformations? In fact, Ferkiss maintains that automation and information-processing affect neither the centralization of capitalist power, the class system, the alienation of labor, the masses' lack of access to information, the old methods of teaching, nor urban violence. Concluding that technological determinism is doomed to failure, Ferkiss borrows from Heilbroner the notion of ideological inertia: the old industrial system holds fast. Because it has not articulated a single ethical principle, a single new value, the "second industrial revolution" has brought only social chaos and economic confusion. The creation of a "technological man" must not be abandoned to the technologists. According to Ferkiss, a true existential revolution would require the following ingredients: 1) a new naturalism, reuniting technological man with nature; 2) a new "totalism," replacing our mechanistic concepts with the idea of system: spirit-body-society-nature unified; 3) a new immanentism, locating the universe "within and not without" according to a logic that Edgar Morin would call "morphogenetic." Rather than seek a symbiosis with machines, man would strive to distinguish himself from them by exploring his inner space. Cultural changes would depend upon a political postindustrial revolution—one which, however, Ferkiss only vaguely adumbrates.

1975), p. 141.

[10]Victor C. Ferkiss, *Technological Man* (New York: New American Library, 1969), p. 173.

In *The Future of Technological Civilization* (1974), Ferkiss formulated his position in political terms. Countering traditional liberalism—as well as false conservatism, technophile socialism, and despairing radicalism—in the United States, the book is a lengthy plea for an ecological humanism. Creating and exerting political control over his technology, the new "technological man" would mark a new stage in the evolution of culture. The political philosophy of the United States—the foundation of the capitalist piracy against nature—may be the anti-naturalism of Locke. But, according to Ferkiss, the socialist world is not much better off with Marx from an ecological standpoint. Both the Lockean and Marxian philosophies correspond to out-of-date scientific paradigms. Ferkiss's eco-humanism would be grounded in "the nature of nature," an expression taken up by Edgar Morin as the first element in his methodological triad.[11] A fount of values, nature becomes normative: it is a preserver, negentropic and limiting. Respect for these limits and for life itself would produce a new scientific system of values. Its poets might be Thomas Merton and Gary Snyder, its moralist the conservationist Aldo Leopold. . .

III. *Technology and the Literary Mind*

Technology is not science. Is it even *applied* science? I shall not remind you that the steam engine was invented before thermodynamic theory could account for it. Technology can and does generate scientific knowledge. But its use of knowledge seems inseparable from big money, big power, big disasters. Scientists complain that they always have to clean up after technologists. Literature has bound, unbound, and rebound Prometheus on his smelly rock, found Faust guilty of all sins except for a fine excess, and can itself be accused of harboring a bias in favor of science and against technology.

The contrasting of technological and literary imaginations, however, is too often and too easily reduced to the dichotomy of technological mastery versus literary knowledge. When Michel Butor, for example, sets out to deny psychological depth, and to practice a certain art of collage or *bricolage*, an art of mosaic intent on surfaces alone, is his goal knowledge or mastery? Jean-François Lyotard has suggested that the technology of assembling *Mobile: Etude pour une représentation des Etats-Unis* is already an urban technology. There is, thus, a congruence between the forms of contemporary fiction and the art and technology of our civilization. Fiction, art, technology wield a big cutter. "The world was not cut right, let us recut it, make it more beautiful. . . ." The destruction of the world by the artist as technologist has a motive hidden in the ambiguity of the

[11]Edgar Morin, *La Méthode: I. La Nature de la nature* (Paris: Editions du Seuil, 1977).

word *mobile,* as in automobile and in motivation, *mobile du crime*. That hidden motive is to replace the power of the world, which began without humans and will probably end without them, by the power of the human ego, which is unable to sustain the thought of being in excess in the world. If literature cannot escape the dilemma that to improve the world is only to make it worse, it still has "a choice of technologies" in the sense that one says "a choice of weapons." But do those who live out there, in a world seemingly dominated by giant machines, also have a choice? The question is: How can the literary mind transform its power from a choice of words into a choice of world?

Benamou planned to conjoin his codification and exploration of "the four humors of discourse about technology" with an examination of "the response of literature to the effects of massive technological change on human affairs." Although the passages dealing with the literary mind as a manifestation of the technological imagination are the most fragmentary of Benamou's working papers, he had noted his basic premise:

> *The problem in the relationship of technology and the literary mind is not the onus of defining technology. Everybody knows what technology means: the urban-industrial context of our lives. The problem is to describe with enough precision the relatedness of the literary mind to technology* defined as *the context of our lives—as a mind-set that pervades the thinking of most urbanized Westerners.*

He intended to define this "mind-set" in two steps. He would first recapitulate—with references to Bachelard, Poulet, and Foucault—a history of circle and grid metaphors. He would then, it appears, have argued that the traditional metaphors both favored by and used to describe pre-scientific and scientific imaginations—the circle and the grid—are superimposed and internally complicated when used by and used to describe the technological imagination. Thus, another binary opposition proliferates into differential relations similar to those of his map of discourses.

Benamou noted that:

> *The relationship between technology and literature includes the fact that technology, like literature, represents a way of knowing the world—we called it a mind-set—and that, like literature, it favors certain forms.*
>
> *In recent art—especially postmodern literature, which reflects grid-like forms of technological style—only the grid may be more akin to a printed circuit—the squaring of the circle emerges as a complementarity between content (the attitude toward technology) and form (the artistic response to the problematic medium or media).*

We can see the punning connection he was making between technocriticism and the literary mind: the circle and the grid—representing poetic holism and scientific objectivity—do not, in his view, apply to the technological imagination. The binary formulation becomes an oxymoron: the "grid" of the "circuit." And since the discourse about technology cannot pretend to objectivity, it invites, like literature influenced by technology, analysis of its preferred forms and metaphors.

Benamou did not intend, however, to analyze metaphors as such. As Leo Marx has shown, the circle and the grid have also been traditionally associated in American literature with nature and the city. Benamou intended to explore a recent literary metaphor for the city—the grid-like circuit—as reflective of a technological "mind-set." Since technocriticism also concerns itself with urbanization, this metaphor would have provided a link between the literary mind and technocritical thought.

Notes concerning "happy technophiles" suggest that he planned to explore that link with reference to Butor's Mobile*—his grid-like mapping of words as cities—within the terms of Lyotard's reading of Butor.*[12] *Thus, from John Cage and Apollinaire—whose* Calligrammes*, in the Gallimard edition, are introduced by Butor—Benamou was moving toward Lyotard's analysis of the "biblioclastic" tradition to which all three belong. "Biblioclasts, like iconoclasts," Benamou noted, "destroy traditional elements of the book in order to create a new medium which resonates with a technological environment." By relating this conjunction of medium and environment to one urban vision—Paolo Soleri's "arcologies"—Benamou would have made his connection between technocriticism and the literary mind at the "happy technophile" corner: "Not only do forms of fiction undergo electronic transformation, but the image of 'urban man' is metamorphosed by happy technophiles."*

Notes concerning "desperate technophobes" similarly connect technocritical and literary visions of the city. From Jacques Ellul's The Technological Society*, with its "grim account of Paris in the 1950s" and its vision of "urbanitis," Benamou was moving toward the "surrealist Marcuse" of* Essay on Liberation*: "From being a desperate technophobe in* One-Dimensional Man*, Marcuse became a hopeful one. Phobia means fear, but fear is not incompatible with hope. A return to basic aspirations, as expressed in Surrealist poetry, still seemed possible, for reasons involving the stability of biological drives." Subsequent notes refer this hope to "the Surrealist beautification of Paris as a city of parks"—a beautification that Benamou apparently found in André Breton's* Poisson soluble.

Benamou's working papers leave off at this point. It is, perhaps, fitting that they do, for there is something of the surrealist gesture in his having planned an essay suspicious of the "science of systems" around an elaborate map of discourses—a map that wants to square its own circle. The gesture itself suggests, in its playfulness, Benamou's faith in the imagination's ability to "transform its power from a choice of words into a choice of world."

Edited by Charles Caramello

[12]See Jean-François Lyotard, *Rudiments païens* (Paris: 10/18, 1977), pp. 81–114.

II
Machines, Myths, and Marxism

Introduction

ANDREAS HUYSSEN

Toward the end of the stabilization phase of the Weimar Republic in 1929, Bertolt Brecht published a satirical poem called "700 Intellectuals Worshipping an Oiltank." In this poem, Brecht attacked the cult of technology which was rampant among the writers and artists of the *Neue Sachlichkeit* (New Objectivity) in Germany, and was indeed widespread in most industrialized countries. Of course Brecht's concern was not that of the conservative culture critics who were bent on salvaging high culture from the encroachments of technology, urbanization, and massification. Brecht did not attack the intellectuals for taking an interest in the oiltank. His satire aimed rather at the transformation of the oiltank into an object of worship and mystification. Thus it serves as a critique of any writing, utopian or dystopian, that treats technology as autonomous, governed not by social forces, but rather by inherent uncontrollable laws. To people holding this view, technology does not represent the most advanced stage of human mastery over nature; instead it assumes the mystical quality of a "second nature" which in turn controls all social life. Of course the roots of this autonomous technology tradition reach way back into the past, in the United States as well as in Europe. But it was in the 1920s that technology first assumed a major ideological role in legitimizing bourgeois domination, a role which—due to changes of the capitalist economy in the wake of increasing monopolization, taylorization, and state intervention—

could no longer be fulfilled by the liberal myth of the "free market" alone. It was precisely this ideological function of the technology cult of the 1920s and its literary manifestations which Brecht attacked.

This critique by the left also informed large segments of the historical avant-garde from Dada to World War II; it was combined with a radical onslaught on the tradition of autonomous art which, since the mid-nineteenth century, had been one of the pillars of bourgeois cultural hegemony. Again Brecht is a case in point. In his discussion of the relationship between literary narrative and film, Brecht insisted that the viewer of films will develop new ways of reading narratives and, conversely, that the contemporary writer of narratives is also a viewer of films. Convinced as he was of the inevitable transformation of traditional genres as a result of the emergence of new media, Brecht concluded: "The technification of literary production is irreversible." Brecht welcomed the invasion of media technology into the sacrosanct sphere of "high" literature, and his own theory and practice of the learning play and the epic theater must be understood as a response to both the social and technological changes in the sphere of cultural production and consumption.

The key question here is not so much to what extent the development of modern technology fuelled the artists' imagination, i.e., to what extent technology became the content of the art work. The point is to understand how technology penetrated to the core of the work itself, dissolving the myths of autonomy, of uniqueness, and of the organic nature of the artistic creation—myths which have prevailed in traditional views of high culture. The invasion of technology into the very fabric of the art object can best be grasped in artistic and literary practices such as collage, assemblage, and montage; it finds its ultimate fulfillment in photography and film, art forms which can not only be reproduced like any other, but are actually designed for reproducibility. In his famous essay "The Work of Art in the Age of Mechanical Reproduction," Walter Benjamin argued that this mechanical reproducibility had destroyed the traditional art work's aura and had changed the very conditions of producing, distributing, and receiving art altogether, thus providing the prerequisite for the development of a socialist mass culture. Both Benjamin and Brecht shared the traditional Marxist view that the development of the productive forces—in art as in society at large—is potentially emancipatory and that technology would reveal its inherently progressive thrust once it had shaken off its capitalist fetters. At the same time they were well aware that the productive forces would not automatically explode oppressive production relations, and that advanced technologies such as radio and film could actually be used to support the existing system of domination and to prevent social change. Neither Brecht nor Benjamin ascribed a progressive function to literary techniques (*Verfremdungseffekt*) or media technology alone. They believed, however, that new artistic techniques and the new media were needed to create a culture in which the traditional gap between high and low, elite and popular would be abolished. The ultimate criterion for success or failure was no longer the completed work as a fetishized object, but rather its reception, its capability of generating new modes of perception, its active place in the historical process. It is in line with this definition of art as process and with Brecht's and Benjamin's understanding of the changing function of intellectuals that they conceptualized the artist not as creator, but as producer and technician. As producer the artist was to do more than just deliver his works to the media apparatus which then would market and distribute them. The production of critical art was not sufficient unless accompanied by the attempt of the artist-producer to change the apparatus of publication and media itself.

To be sure, Benjamin's belief in the revolutionary effect of media technologies and reproduction techniques in art and literature has not withstood the test of time. The specific use that both German fascism and the American culture industry have made of the mass media proves beyond doubt that the hopes for the emancipatory potential of media technologies were too high flown. And it was precisely this experience of fascism and the American culture industry of the 1940s and 1950s that is at the root of Adorno's blanket condemnation of the culture industry as mass-deception. For Adorno, the culture industry is nothing more than one aspect of that dialectic of enlightenment in which technical rationality has become the rationality of domination *per se,* and thus the notion of technology becomes the center stone for the theory that art and artistic technique are irreconcilable with mass mediated culture. As a matter of fact, Adorno radically distinguished between the concept of technique in the culture industry and technique in works of art. He held that in the latter, technique is concerned with the internal organization of the object itself, mediated as it may be with social processes, while technique in the culture industry remains entirely external to its object. He even went so far as to say that ultimately the technification of the art work aimed at its abolition. By holding on to the essential autonomy of the work of art in face of universal manipulation and deception, Adorno excluded a priori the potential emergence of a new kind of art in which aesthetic form, political content, and technical reproduction could be successfully mediated.

The debate between Adorno on one side and Brecht and Benjamin on the other, sketched here in its barest outline, is somewhat dated since the positions taken responded to specific historical experiences and were based on different philosophies of history. While Adorno was drawing on his experience of fascism, Brecht and Benjamin reacted to developments in the Weimar Republic and in the early years of the Soviet Union. The respective historical differences notwithstanding, these theories and the struggles which generated them remain crucial to discussions of literature and technology in a country where the achievements of the pre-World War II avant-garde have been heavily distorted by being subjected mainly to formalist and intrinsic approaches.

To be sure, Brecht, Benjamin, and Adorno represent a single, though crucial constellation in a wide spectrum extending from the Dadaists' iconoclastic use of technology to the Russian post-1917 avant-garde, its productivism and proletcult, and finally to the French surrealists' attempt to explode the reifications of an increasingly technologized life world. Summing up then, two basic assumptions can be made for the analysis of any of these movements, their successes and their failures:

1) Changes in technological production bring about changes in the everyday life world which then are taken up (reflected, represented, reutilized, transformed) in art and literature. More importantly: these changes in technology transform the technological structure of art production itself;

2) Technological progress—in art or industry—is never completely identical with the way in which new technologies are made to function. Every technological advance potentially holds a wide span of realizations, but this potential is circumscribed by the strength or weakness of the dominant relations of social production. Consequently, this span is not per se open to social change, nor, however, is it ever totally closed.

Historical developments since the 1920s have made it mandatory to work toward a critical theory of technology which transcends the orthodox Marxist belief in technology's emancipatory power and which at the same time steers clear of any demonization or worship of technology as an uncontrollable force. It is this critical perspective that informs the four essays assembled in this section. The theoretical piece by James Schmidt and James Miller critiques Marx's views of technology as a neutral tool. Schmidt and Miller show that the roots of Marx's uncritical acceptance of technology are to be found not so much in nineteenth-century scientism, but in his indebtedness to the post-Renaissance tradition of Western humanism and enlightenment. What is placed in question here is not only a specific tradition of technological optimism and determinism within Marxism, but indeed the whole Western ideology of progress and efficiency which Rousseau attacked so eloquently in the eighteenth century. Schmidt and Miller's use of Rousseau points to the possibility and necessity of reexamining that long literary tradition of anti-technological romanticism and romantic anti-capitalism, to unlock its critical moments and to mediate them with the critique of reification which has been the central project of critical Marxism since the 1920s.

Helen Fehervary shares Schmidt/Miller's critique of automatic technologism and reification and shows how a serious critique of technological progress has emerged within the cultural sphere of the German Democratic Republic (GDR), i.e., within Marxism itself. The analysis of technology as a legitimizing ideology in East European state socialism is at the center of Fehervary's essay. She examines variations on the Prometheus myth in Marx and in the writings of a new generation of East German poets, playwrights, and novelists, some of whom are clearly pursuing the Brechtian project under changed historical conditions. Only now the critique is aimed at socialist technocrats, not at bourgeois intellectuals. In short, the critique of "socialist technology" turns into a scathing critique of a form of domination in which the myth of technological progress has obliterated the Marxian dialectic of Prometheus as rebel and as producer.

David Bathrick discusses this same dialectic in the context of the left avant-gardes of the early years of the Soviet Union and focuses on its implications for cultural policy in the Soviet Union and East Germany. He shows how the productivist aesthetic of the post-revolutionary period, which held a tremendous emancipatory promise, actually feeds into official cultural policy and is no longer viable for contemporary writers. His thesis is that major East German writers today are genuine avant-gardists precisely because they reject the "classical" Soviet and Brechtian tradition of the 1920s.

Jost Hermand's concluding essay on Ernst Jünger offers insights into the technology cult of the extreme right between the two World Wars. Here the mystification of technology goes far beyond the enthusiasm of the liberal representatives of *Neue Sachlichkeit* whom Brecht attacked. The object of worship is no longer the oiltank of the thriving Weimar Republic in the process of modernization, but the war machine, technology at its most destructive. This aesthetization of war is not a peculiarly German phenomenon; it had a precursor in Italian futurism, especially in Marinetti's manifesto on the Ethiopian war. Even today it remains important to lay open the ideological presuppositions of technophiles such as Jünger, whose renewed popularity points to the undiminished need of capitalism to legitimize technology at a time when the very principle of "progress through technology" is increasingly being challenged.

Aspects of Technology in Marx and Rousseau

JAMES SCHMIDT and JAMES MILLER

Jean-Jacques Rousseau has always bothered Marxists. Over two-hundred years ago, on the eve of Europe's industrial revolution, Rousseau took a jaundiced view of the "progress" afforded by scientific enlightenment and the technical conquest of nature:

> When, on the one hand, one considers the vast labors of men, so many sciences fathomed, so many arts invented, and so many forces employed, chasms filled, mountains razed, rocks broken, rivers made navigable, land cleared, lakes dug out, swamps drained, enormous buildings raised upon the earth, the sea covered with ships and sailors; and when, on the other hand, one searches with a little meditation for the true advantages which have resulted from all this for the happiness of the human species, one cannot fail to be struck by the astounding disproportion prevailing between these things, and to deplore man's blindness, which, to feed his foolish pride and an indefinable vain admiration for himself, makes him run avidly after all the miseries of which he is susceptible, and which beneficent nature had taken care to keep him from.[1]

There have always been readers who have felt that Rousseau's doubts about progress were pernicious. In acknowledging a complimentary copy of the *Second Discourse,* Voltaire congratulated Rousseau for his "new

[1]Jean-Jacques Rousseau, "Discourse on the Origins and Foundations of Inequality," in his *The First and Second Discourses,* trans. Roger D. Masters (New York: St. Martin's Press, 1964), p. 193.

book against the human race," and suggested that anyone convinced by Rousseau's argument might wish to walk on all fours.

Even today, good Marxists deplore Rousseau as a petit-bourgeois anachronism who, to quote Lucio Colletti, "did not even remotely perceive the problem of development." Serge Mallet, the French theorist of a "new working class," is even more blunt: "The labor movement ought firmly to deplore the pessimistic tendencies which tend to confuse the working class's aspirations to a higher standard of living and to an acceleration of the technical and scientific process with a 'subjective integration into the capitalist system'. . . . There are some stale odors of an anti-technical 'Rousseauism' presently floating in leftist intellectual milieux which are particularly dangerous."[2]

Why this fear of creeping Rousseauism? What did Rousseau say that so angers so many right-thinking progressive intellectuals?

Perhaps it is Rousseau's penchant for posing dilemmas that demand moral choice rather than economic development. For Rousseau, the conflicts within modern society can never be genuinely resolved through technical conquest or the spread of industry. Instead, these conflicts generate an inescapable decision: Do we want a wealth of wants—what Rousseau, following the ancients, called "luxury"—at the cost of envy, selfishness, and social conflict? Or do we prefer the possibilities for virtue and happiness afforded by autonomy, altruism, and harmonious co-existence, at the price of placing *limits* on our wants?

For Marxists, by contrast, the road to social justice is paved with material goods. In our own industrial age, the argument goes, technology and economic growth provide the basis for a true emancipation of men. To those nervous luddites misled by Rousseau or frightened by machines, the progressive revolutionary responds with a lesson in clear logic: the means of production—the factories, assembly lines, computers, and machines—are all only *tools.* Their use depends on who uses them: and the tools which today serve the interests of capitalist exploitation tomorrow can serve the cause of human freedom.

Not all Marxists, however, have been convinced by this logic. Ever since Stalin's forced industrialization of Russia, there have been misgivings about the apparent effectiveness of Marxism as a credo legitimating the human costs of economic development. Such "uses" of the theory have in turn stimulated doubts among those loyal to Marx's program for human emancipation.[3] What *is* the relationship between technology,

[2]Lucio Colletti, *From Rousseau to Lenin,* trans. John Merrington and Judith White (New York: Monthly Review, 1972), p. 157; Serge Mallet, *Essays on the New Working Class,* trans. Dick Howard and Dean Savage (St. Louis: Telos Press, 1975), p. 120.

[3]Lenin's evaluation of the Taylor system is symptomatic: under capitalism it oppresses the worker, under socialism it streamlines production and shortens the necessary labor

economic development, and human freedom? Does economic growth in fact obviate the kind of moral decisions Rousseau thought were inescapable? Does liberation from hunger necessarily augment the autonomy of the individual? Can man always master the machines he has made?

Marx himself had posed the questions of technology and development in "profane" terms: he believed that social individuals capable of mastering the tools of modern industry and putting them to humane use were educated naturally through the labor process and class conflict characteristic of civil society. But to grasp *how* Marx himself understood the applied technology of modern industry, and *why* he assumed it would further the cause of human freedom, it is necessary to grasp beforehand the *terms* in which Marx approached the problem: for these terms already imply an answer to the problem they address. The key terms here are *civil society*, the *social individual*, and the *tool*.

For Marx as for Hegel, civil society is "the totality of the material conditions of life." In the tradition of political thought running from Aristotle to Rousseau, the locus of freedom within civil society had been *politics*—that public space removed from the necessity of labor where citizens discussed issues of common concern and, in the process, received a civic education in moral well-doing. This tradition underwent a decisive transformation in the hands of Hegel, the first political philosopher to incorporate the new theories elaborated by political economists like Adam Smith. In light of their findings, Hegel strictly defined "civil society" to denote exclusively what we today would call economic and social relations: civil society represented that public space where private individuals pursued material well-being through the free exchange of goods produced by labor.

Marx inherited and extended Hegel's understanding of civil society. At the same time, however, both Marx and Hegel maintained the classical

---

time. V. I. Lenin, "The Immediate Tasks of the Soviet Government," *Collected Works*, Vol. 27 (Moscow: Foreign Languages Pub. House, 1965), p. 258; "The Taylor System—Man's Enslavement to the Machine," *Collected Works*, Vol. 20 (Moscow: Foreign Languages Pub. House, 1964), p. 153. Bernard Gendron, *Technology and the Human Condition* (New York: St. Martin's Press, 1977), unintentionally provides a caricature of the orthodox Marxist argument. Among the more important criticisms of Marx's views on technology are Jürgen Habermas, "Science and Technology as 'Ideology'," in *Towards a Rational Society*, trans. Jeremy Shapiro (Boston: Beacon Press, 1970), pp. 81 ff.; Habermas, *Knowledge and Human Interests*, trans. Jeremy Shapiro (Boston: Beacon Press, 1971); Albrecht Wellmer, *Critical Theory of Society*, trans. John Cumming (New York: Herder and Herder, 1971), pp. 67–120; and Jean Baudrillard, *The Mirror of Production*, trans. Mark Poster (St. Louis: Telos Press, 1975). While we concur with much of this criticism, we disagree with the interpretation of the source of the problem in Marx. Rather than attributing Marx's inability to criticize technology adquately to a "latent positivism," or a penchant for "instrumental rationality," or "productivism," we locate the problem in his debts to Enlightenment humanism, in particular to his acceptance of the principle of "*verum factum*." See our discussion below.

connection between civil society and civic education. Although civil society consisted of market transactions and labor to satisfy needs, Hegel argued that these instrumental activities forced the individual to reflect on his actions from a universal point of view.[4] Similarly Marx: although he denounced the narrow selfishness engendered by market transactions, Marx also found a progressive pedagogical aspect of civil society in the workers it produced. By redescribing civil society as the locus of an enlightening and epochal class conflict, and by praising the knowledge gained in the labor process proper, Marx was able to preserve classical hopes for human perfectibility on the novel basis of a civic education gained, not through the display of moral character in political discourse, but rather through the dissolution of social domination in class struggle and the mastery of nature by industry.

In the process, Marx replaced the traditional picture of the cultivated individual as a creature of defined skills and moderate desires with a new image, depicting the free individual as a sensuous and practical being, fit for a wide range of tasks and entertaining a rich variety of wants. Where a thinker like Rousseau found the perfection of man in a closed code of moral integrity and material self-sufficiency, Marx exults in the open dynamism of the human being: indeed, it is precisely the ability to posit and realize ends that Marx hopes a communist society will liberate.

The novelty as well as the attractiveness of Marx's image is most readily apparent in the *Grundrisse,* where he waxes lyrical about the *social individual* cultivated within civil society:

> When the narrow bourgeois form has been peeled away, what is wealth, if not the universality of wants, capacities, enjoyments, productive powers, etc., of individuals produced in universal exchange? What, if not the full development of human control over the forces of nature—those of his own nature as well as those of so-called 'nature'? What, if not the absolute elaboration of his creative dispositions, without any preconditions other than antecedent historical evolution which makes the totality of this evolution—i.e., the evolution of all human powers as such, unmeasured by any *previously established* yardstick—an end in itself? What is this, if not a situation where man does not reproduce himself in any determined form, but produces his totality? Where he does not seek to remain something formed by the past, but is in the absolute movement of becoming.[5]

[4]For Marx, see Karl Marx, *A Contribution to the Critique of Political Economy* (Moscow: Foreign Languages Pub. House, 1970), p. 20. On the transformation of the concept of "civil society" see Manfred Riedel, "Hegels Begriff der 'Bürgerlichen Gesellschaft' und das Problem seines geschichtlichen Ursprungs," in *Studien zu Hegels Rechtsphilosophie* (Frankfurt am Main: Suhrkamp, 1969), pp. 135ff. For Hegel's views, see G. W. F. Hegel, *Philosophy of Right,* trans. T. M. Know (New York: Oxford Univ. Press, 1952), esp. para. 187, 197.

[5]Karl Marx, *Grundrisse: Foundations of the Critique of Political Economy,* trans. Martin Nicolaus (New York: Random House, 1973), p. 488.

This is heady stuff: Everyman a Prometheus, no less. Moreover, Marx at the close of this passage makes perfectly clear what has been left behind: "the childish world of antiquity," with its "satisfaction from a limited standpoint." By setting the enormous productive power of industry to work elaborating new human wants, Marx anticipated the perfection of a multi-faceted "social individual" uniquely equipped by previous experience to master the new industrial forces.

In this evaluation of the liberatory potential of science and technics, Marx treats modern industry as a clear-cut extension of man's productive power: he essentially approaches it as a complex and potent development of the *tool*.

In *Capital* Marx defines the tool as "a thing, or complex of things, which the worker interposes between himself and the object of his labor, and which serves as a conductor, directing his activity onto that object."[6] In this definition, the tool unambiguously augments human agency: Marx in *Capital* therefore endorses Benjamin Franklin's view that man is a "tool-making animal," and echoes a remark of Hegel's in the *Science of Logic:* "In his tools, man possesses power over external nature."[7] This view implies the "neutrality" of technics: the tool only expands existing human powers, it does not qualitatively transform them. Moreover, the use to which a tool is put depends on the ends of the craftsman who uses it; it can only extend, not limit, the scope of his agency.

When Marx comes to discuss the *machine* in *Capital,* he merely enlarges his definition of the tool: "The machine . . . is a mechanism that, after being set in motion, performs with its tools the same operation as the worker formerly did with similar tools."[8] On the other hand, *machinery*—that is, an organized system of machines such as those employed on a modern assembly line—at first glance involves something fundamentally new. A "motive mechanism" with an "independent form, entirely emancipated from the restraints of human strength," machinery looms as a "mechanical monster whose body fills whole factories, and whose demonic power, at first hidden by the slow and measured motions of its gigantic members, finally bursts forth in the fast and feverish whirl of its countless working organs."[9]

Marx's imagery here implies that modern machinery accomplishes a qualitative metamorphosis in the relation between human action and its implements. Elsewhere in *Capital,* Marx describes how "large-scale

[6]Karl Marx, *Capital: Volume I,* trans. Ben Fowkes (New York: Random House, 1977), p. 285.

[7]G. W. F. Hegel, *Science of Logic,* Vol. II, trans. W. H. Johnston and L. G. Struthers (London: G. Allen and Unwin, 1966), p. 388.

[8]Marx, *Capital I,* p. 495.

[9]*Ibid,* p. 503.

industry" confronts the worker as a "pre-existing material condition of production," an alien force over which the worker himself has no control. By the very scale and complexity of its technology, modern industry necessitates a specific organization of society: namely, a coordinated collective labor force, trained to man the new engines of industry with their "machines of Cyclopean dimensions."[10]

Despite this ominous imagery, and its implications, Marx never fails to defend modern industry as "revolutionary"; it continuously transforms not only "the technical basis of production but also the functions of the worker."[11] Marx sees in this transformation of technique and function the real basis on which emancipation rests. True, within the current capitalist relations of production, the factory system and the vagaries of the economy toss workers helplessly from city to city, job to job. But the very "fluidity of functions and mobility" within civil society produce a new man, the social individual, formed by the imperatives of social production and uniquely capable of mastering the wayward tools of modern industry.

Indeed, when all is said and done, it sounds very much as if for Marx the most complex machinery is simply the tool writ large. Once the oppressive capitalist shell has fallen away, Marx foresees the emergence of the "totally developed individual, for whom the different social functions" demanded by modern industry "are different modes of activity he takes up in turn."[12] The awesome powers of industry, which formerly appeared alien, now merely indicate "to what degree general social knowledge has become a *direct force of production,* and to what degree, hence, the conditions of the process of social life have come under the control of the general intellect and been transformed in accordance with it."[13] Cyclopean machines become obedient servants before the social individuals, like Marx's new Renaissance men.

There are clear echoes here of the sixteenth century mechanical arts movement which in fact helped launch the new age of industry and science. The "new men" of the Italian Renaissance, artisans and mechanics outside the university, argued that knowledge—*true* knowledge, capable of proving itself by producing good works—came, not from the interpretation of classic authors, but instead from the construction of machines: hence their principle, *verum factum,* 'truth through doing.' Their understanding was eventually given philosophical coherence by Bacon and Descartes, who speculated that the world was a vast machine created by God. Our knowledge of this world could proceed either by

[10]*Ibid,* p. 508.
[11]*Ibid,* p. 617.
[12]*Ibid,* p. 618.
[13]Marx, *Grundrisse,* p. 706.

building miniature replicas of it in laboratories, or by analyzing and dissecting parts of the machine itself.[14]

Of course, God may outwit us: as Bacon put it in one of his less aggressive moments, nature plays hide and seek. But there is one domain in which a transparent understanding is feasible: in those fields where man himself has fabricated the objects under study. This was the lesson Vico drew from Hobbes: "In the night of thick darkness enveloping the earliest antiquity, so remote from ourselves, there shines the eternal and never failing light of a truth beyond all question: that the world of civil society has certainly been made by men, and that its principles are therefore to be found within the modification of our own human mind."[15]

From this angle, it only seems natural that Marx would conclude that technology, which served as the model for *verum factum* in the first place, presented no insoluble problems to human understanding and mastery: as Marx remarked in the *Grundrisse,* "nature builds no machines."[16] Comparing the prospects for a history of technology with the "history of natural technology" Darwin had written, Marx asked: "Does not the history of the productive organs of man in society, of organs that are the material basis of every particular organization of society, deserve equal attention? And would not such a history be easier to compile, since, as Vico says, human history differs from natural history in that we have made the former, but not the latter?"[17] Technology for Marx shows us the "active relation of man to nature," while the history of industry gives us "the *open* book of man's essential powers."[18] These phrases, like Marx's favorite maxim, "nothing human is foreign to me" (*Nihil humanum a me alienum puto*), reflect his debt to Vico, and an earlier tradition of "practical" humanism.

They also suggest that Marx's "humanism" is not, as often assumed, a check on his trust in the emancipatory potential of technology and economic development. On the contrary, Marx's humanism helps explain his ultimate ability to approach technology critically. After all, *if* world history resembles a man-made machine with a handle, *if* technologies are only *tools,* and *if* civil society generates *social individuals* equipped

[14]For an overview of this period see Paolo Rossi, *Philosophy, Technology and the Arts in the Early Modern Era,* trans. Salvator Attansio (New York: Harper & Row, 1970), esp. pp. 137–45.

[15]*The New Science of Giambattista Vico,* trans. Thomas Goddard Bergin and Max Harold Fisch (Ithaca: Cornell Univ. Press, 1948), para. 331. For Hobbes, see *De homine* X:5.

[16]Marx, *Grundrisse,* p. 706.

[17]Marx, *Capital I,* p. 493n.

[18]*Ibid.*; Marx, *Economic and Philosophic Manuscripts of 1844* in Karl Marx & Frederick Engels, *Collected Works,* Vol. 3 (New York: International Pub. Co., 1975), p. 302.

with the "general intellect" to take hold of this machine and these tools, *then* indeed human emancipation can occur as Marx anticipated, through the recovery of the essential human powers embodied in industry, but currently alienated and under the domination of capital.

But is history a machine? Are technologies tools? And does civil society in fact equip workers with the knowledge necessary to master the machines other men have made?

If we are going to use machine metaphors, we might as well consider a modern machine. For example, where is the handle on a computer? Consider this statement by Marvin Minsky, a prominent researcher in the field of "artificial intelligence":

> When a [computer] program grows in power by an evolution of partially understood patches and fixes, the programmer begins to lose track of internal details, loses his ability to predict what will happen, begins to hope instead of know, and watches the results as though the program were an individual whose range of behavior is uncertain. . . . Large heuristic programs will be developed and modified by several programmers. . . . The programs will grow in effectiveness, but no one of the programmers will understand it all. (Of course, this [attempt to expand the effectiveness of the program] won't always be successful—the interactions might make it get worse, and no one might be able to fix it again!) Now we see the real trouble with statements like 'it only does what its programmer told it to do.' There isn't any one programmer.[19]

As Joseph Weizenbaum, one of the founding fathers of computer science, has remarked about Minsky's statement: "If a program has outrun the understanding of the agents who created it, what can it mean for it to 'grow in effectiveness,' or for that matter, to 'get worse'?"[20]

Faced with such technologies, the idea that they are simply neutral tools becomes, as one critic puts it, a "bromide." It permits us to retain the illusion that we can redirect technical means to whatever ends we choose. But this "is to see complex technologies as if they were handsaws or egg beaters." We do not *use* many modern technologies so much "as live them"; they are "tools without handles."[21]

[19]Marvin M. Minsky, "Why Programming is a Good Medium for Expressing Poorly Understood and Sloppily Formulated Ideas," as quoted in Joseph Weizenbaum, *Computer Power and Human Reason: From Judgement to Calculation* (San Francisco: W. H. Freeman, 1976), p. 235. Weizenbaum goes on to note (p. 238) that target selection programs used in Pentagon computers during the American War against Viet Nam approximated the prospects outlined by Minsky. Programmed by individuals who were no longer present, the computers were used by individuals who had not the slightest idea what specific criteria the programs used to select "free fire" zones in Viet Nam.

[20]Weizenbaum, p. 235.

[21]Langdon Winner, *Autonomous Technology: Technics Out of Control as a Theme in Political Thought* (Cambridge, Mass.: MIT Press, 1977), pp. 27, 200, 202.

The inapplicability of the tool metaphor and the questionability of the principle *verum factum* are significant. They force us to take seriously the alternatives Rousseau so starkly posed: wealth *or* autonomy, luxury *or* harmonious coexistence. For Rousseau, an abundance of material goods by no means guaranteed that life would become any more meaningful, or that the individual would extend his sphere of autonomous freedom; indeed, Rousseau feared that human beings who develop the knowledge, skills, and desires necessary to subdue nature and amass wealth might well be corrupted by the experience. Far from undergoing an education in practical freedom within civil society, they may discover instead that ideas and sentiments essential to their harmonious coexistence have been lost, since their particular wants have become limitless, and the complexity of their society has made their understanding ineffective, these individuals, unable to comprehend the common good, will no longer care for the restraints imposed by fellow-feeling, or the sacrifice entailed by mutually-binding norms.

Rousseau's skepticism about the "general intellect" embodied in applied technology and modern science was shared by Max Weber, who argued that science has done little to increase the individual's ability to understand or control his world. Have "we, today . . . sitting in this hall . . . a greater knowledge of the conditions of life under which we exist than has an American Indian or Hottentot?" asked Weber. "Hardly. Unless he is a physicist, one who rides in the streetcar has no idea how the car happened to get into motion. And he does not need to know. He is satisfied that he may 'count' on the behavior of the streetcar, and he orients his conduct according to this expectation. . . . The savage knows incomparably more about his tools."[22]

Weber elsewhere raises the issue Rousseau had previously broached in a different context: the issue of providing some normative code to secure meaning in life, and to provide criteria for controlling the direction and tempo of change. Once unhinged from such classical norms as virtue and happiness, independence and moderation, the purely technical values of efficiency, productivity, and predictability threaten to create precisely the kind of human beings Weber dreaded, "specialists without spirit, sensualists without heart."[23] And who, looking around today at socialist as well as capitalist countries, can in good conscience defend the validity of Marx's contrary vision? Can we yet see the glimmerings of those new Renaissance men Marx felt confident would emerge from the factories constituting modern industrial society?

[22]Max Weber, "Science as a Vocation," in *From Max Weber,* trans. H. H. Gerth and C. Wright Mills (New York: Oxford Univ. Press, 1946), p. 139.

[23]Max Weber, *The Protestant Ethic and the Spirit of Capitalism,* trans. Talcott Parsons (New York: Scribner, 1958), p. 182.

We do not propose that men walk on all fours. But once it is granted that applied science and an abundance of material goods do not automatically increase human freedom and happiness, we are forced to evaluate technologies and economic development in a different light. A critical perspective would discard any general rule commending capital-intensive technology or "growth" as a panacea for social ills. A critical perspective would also raise a series of questions aimed at ascertaining what *kinds* of technology are appropriate to a self-governing cooperative society interested in securing the autonomy of individuals. For example: Does a given factory facilitate the participation of workers in managing and thus understanding the industrial process? Is a given technique of assembly amenable to varying the work each individual performs, or does it require a fragmented division of labor? Is a given industrial design open to community control, or does it entail extensively reorganizing the community to accommodate the needs of the industry? Are the goods produced by an industry deemed useful by the community? A meaningful discussion of these issues cannot presuppose that technologies are tools, or that technical expertise invariably serves the cause of human freedom; nor can these issues be properly addressed so long as the values of efficiency, productivity, and predictability remain the primary criteria for evaluating technologies.

At times, Marx recognized this; great thinkers are never slaves to their theories. Indeed, the fear of "creeping Rousseauism" among some modern Marxists betrays an uneasy conscience: the force of Marx's imagery when he shows the spell of things over men mocks any bland assurance that these Cyclopes can be readily tamed. By thinking through Marx's bleakest images, a critical theory of technology can still arise out of Marxism, sharing with Marx a commitment to dissolve oppressive social relations, while questioning the specific forms genuine freedom can take. With his characteristic sense for this sort of oblique truth, Walter Benjamin put it this way: "Marx says that revolutions are the locomotives of world history. But the situation may be quite different. Perhaps revolutions are not the train ride, but the human race grabbing for the emergency brake."[24]

[24]Walter Benjamin, *Gesammelte Schriften* I:3, ed. Rolf Tiedemann and Hermann Schweppenhäuser (Frankfurt am Main: Suhrkamp, 1974), p. 1232. Cf. Walter Benjamin, "Theses on the Philosophy of History," in his *Illuminations*, ed. Hannah Arendt, trans. Harry Zohn (New York: Harcourt, Brace & World, 1969), pp. 253–64.

Prometheus Rebound: Technology and the Dialectic of Myth

HELEN FEHERVARY

"Fritz Cremer, Bronze: 'Man Climbing Upward'" is the title of a poem by the East German singer and expatriate Wolf Biermann. This poem was published in 1968, the year that saw the movement toward a self-determined socialist future dramatically thwarted by political and military intervention in Czechoslovakia. It articulated in cultural terms the political-economic contradictions underlying socialism in the Soviet Bloc, and addressed itself to a paradigmatic example of the dominant ideology and of socialist realism: a Promethean figure in bronze created by the most successful sculptor of the German Democratic Republic (GDR), the National Prize winner Fritz Cremer. Ironically, almost ten years later, in November of 1976, Cremer was to become one of the original thirteen signers of the declaration by intellectuals protesting Biermann's expatriation from the GDR.[1] Two days later, claiming he had been misled by the other signers, Cremer withdrew his name from the protest declaration and joined the ranks of those denouncing Biermann on the front pages of the Party organ, *Neues Deutschland*.[2] The poem itself, like the later turn of events which it foreshadows, demonstrates how

[1] The other signers were Christa Wolf, Heiner Müller, Volker Braun, Gerhard Wolf, Sarah Kirsch, Erich Arendt, Stephan Hermlin, Jurek Becker, Günter Kunert, Stefan Heym, Franz Fühmann, and Rolf Schneider. For reports on the events surrounding Biermann's expulsion see *New German Critique* 10 (Winter 1977).

[2] See *Der Spiegel*, 22 Nov. 1976, 30.

fragile the relationship has become between the proletarian positive hero of socialist progress and the patent-leather-shoed technocrat of industrial achievement:

Fritz Cremer, Bronze: "Man Climbing Upward"

Climbing laboriously
Climbing steadily
Climbing irresistibly upward: A man

Man, that's some climb:
He's falling!
He's almost falling!
He does look like he's falling
He almost looks as if he's falling

But climbs up
This guy is climbing
He is really climbing
He's climbing with all he's got!
This guy has eaten old Newton's apple:
He's just climbing

Not yet the clutching toes, but
He's ripped his heel from the ground
Above the knee the tendons stretch the leg
To make it or break it upright
The leg then thrusts into the pelvis
The hips lunge high
Upward also the massive paunch torments its way
Girding the chest
The ligaments flow
Past the head bent mighty
Into the shoulder. Discharge
Into the arm. Gush forth
Into the hand. Shoot up
Into the finger tips. Yes!
This mass of flesh rises up
This colossus climbs and climbs
—this is really a climber!
He climbs irresistibly
—and laboriously too, I said that already—
This man we rightly name:
MAN CLIMBING UPWARD

But just tell us one more thing:

Where is this guy going?

Up there where he's climbing
What's there? Is there really

an up there?

Say, is he climbing up to join us?
Or is he climbing up out of our midst?

Is this guy leading us?
Or is he leaving us?

Is he pursuing somebody?
Or is he fleeing somebody?

Is he making progress?
Or is he making a career?

Or is he perhaps to be—what we've surmised all along:
A symbol of the species man?
Is this being climbing up
To freedom, or—what we've surmised all along:
To the flesh-pots?
Or is mankind climbing up there
In a mushroom cloud to God and—what we've surmised all along:
Into Nothing?

So many questions about a man climbing upward[3]

Biermann's skeptical attitude toward technology and production is symptomatic of all progressive writing in the GDR since the late 1960s, represented by major figures like Christa Wolf, Günter Kunert, Volker Braun, Heiner Müller, and Thomas Brasch. Indeed, it is striking that precisely these writers, who follow most directly in the avant-garde tradition of Expressionism and the Proletkult—of Ernst Bloch, Walter Benjamin, Erwin Piscator, and Bertolt Brecht—should be so critical of a theoretical framework in which their mentors were rooted. The notion that technology could revolutionize culture lies at the center of Piscator's political theater, Brecht's radio theory and his work with theater and film, and Benjamin's theory of the artist as producer and the politicization of art and mass culture through mechanical reproduction. This very optimism toward technology within the critical Marxist tradition of the 1920s and 1930s manifests itself in the GDR today as just that: a naive faith in the instrumentalization of history and human life. Hence "Prometheus Rebound." The key issue here, however, is not technology as such but its larger framework of productivism and instrumental rationality inherited from the Enlightenment.

Let us first backtrack for a moment to Marx, who both came out of and attempted to supersede the Enlightenment tradition. Marx's concept of human labor as the center of life is already present in his reference to

[3]Wolf Biermann, "Fritz Cremer, Bronze: Der Aufsteigende," in his *Mit Marx- und Engelszungen* (Berlin: Verlag Klaus Wagenbach, 1968), pp. 21–22. My translation.

Prometheus in his doctoral dissertation, written from 1839 to 1841: "Philosophy makes no secret of this. The confession of Prometheus: 'In a word, I detest all gods' is her own confession, her own watchword against all the gods of heaven and earth who do not recognize man's self-consciousness as the highest divinity . . . Prometheus is the foremost saint and martyr in the philospher's calendar."[4] In a note to the dissertation he writes: "Just as Prometheus, having stolen fire from heaven, begins to build houses and settle on the earth, so philosophy, having extended itself to comprehend the whole world, turns against the phenomenal world."[5] If Marx addresses himself here specifically to Hegelian philosophy as locating in history the practical moment which overcomes speculative metaphysics, his use of the Prometheus myth also anticipates his later critique of Hegel in *The German Ideology*, the 1844 *Manuscripts*, and the *Feuerbach Theses*, where materialist philosophy is not only to interpret the world but also to change it. For Marx, Prometheus is the archetype of the producer/rebel, at once *homo faber* and man in revolt. Marx extends the bourgeois Enlightenment's notion of progress through material production, and at the same time—especially in his later work—he problematizes the rationalization of production by locating contradictions in the production process itself and by concluding that human beings through labor will also bring about revolutionary change.

In GDR culture this unity of production and revolt, paradigmatic in the Prometheus figure, has become disunified. The figure of the positive hero who dominated GDR culture in the 1950s and early 1960s served primarily as an instrument of legitimation for industrial production and the state, and labor was rationalized as an end in itself rather than being understood as rooted in conflict and change. Prometheus, to continue the metaphor, was coopted; reunited with the gods, as it were, he had no more cause for revolt. Rather than being punished for bringing fire to humans, the gods forgave him, and he recovered his technology to be managed by the gods for the entire universe, heaven and earth: the rational solution to the threat of tragic conflict and historical arrest. This identification of productivism and progress was not limited, however, to the strict party-line literature of socialist realism. Even in the early works of Heiner Müller and Christa Wolf (*The Rate Buster*, 1956; *The Divided Heaven*, 1963), the hero as producer took precedent over the hero as rebel, though Wolf and Müller presented their positive heroes as embodying the ideology of historical necessity. And still today Wolf Biermann, who

[4]*Karl Marx: Early Texts*, trans. and ed. David McLellan (Oxford: Blackwell, 1971), pp. 13–14.

[5]Karl Marx, Friedrich Engels, *Werke, Ergänzungsband* I, ed. Institut für Marxismus-Leninismus beim ZK der SED (Berlin: 1956–68), p. 215; rendered in English in S. S. Prawer, *Karl Marx and World Literature* (Oxford: Clarendon Press, 1976), p. 24.

has been most vocal in exposing the absorption of revolutionary potential by the economic power structure, makes a distinction between the early anti-fascist and reconstruction years and the later technocratic stabilization. In 1953, he recalls, he acquiesced to the Soviet tanks which put down the workers' rebellion, albeit "with tears in his eyes."[6]

As the GDR emerged as one of the world leaders in gross national product in the 1960s, the image of the positive hero disappeared from the works of progressive writers. Christa Wolf's pioneering novel of 1968, *The Quest for Christa T.*, focused precisely on this problem of productivism and paved the way for an entire body of anti-*Bildungsliteratur*, favoring the tradition of modernism to that of the nineteenth-century novel which had been the model for socialist realism. Her heroine, Christa T., is an unproductive anti-hero in every sense of the word. She is asocial, writes fragments that are never published, prefers a contemplative country life to a career, is uncomfortable in the constraints of the traditional family structure, and finally dies of leukemia—of a social imbalance which could not accommodate the questions and alternatives posed by people like her. In direct contrast to the hero of the 1950s, Christa T. is a burden to society. Her labor and goal rest solely in asserting her own place in history, and thereby she disturbs the streamlining of the historical continuum. Her mode of defying instrumentalization in order to become historical subject brings to mind one of Walter Benjamin's notes to his "Theses on the Philosophy of History": "Marx says that revolutions are the locomotives of world history. But the situation may be quite different. Perhaps revolutions are the not train ride, but the human race grabbing for the emergency brake."[7]

In her essay "The Reader and Writer" (1968), Wolf programmatically defends the individual art of writing against the systematization of life through science, technology, and the media. In fact, she attacks Benjamin, albeit the Benjamin of "The Author as Producer" and "The Work of Art in the Age of Mechanical Reproduction," who predicted the end of aura and the collectivization of literary production and reception. Unlike Benjamin, who mechanistically differentiated communism which politicized art from fascism which rendered politics aesthetic, Wolf finds a common denominator in the loss of aura itself and inverts Benjamin's juxtaposition: "If everything depends upon which social order the new

[6]Statement made by Biermann on 13 November 1976 during a concert in Cologne.

[7]An unpublished loose page among the "Theses on the Philosophy of History," quoted by Rolf Tiedemann in the afterword to Walter Benjamin, *Charles Baudelaire. Ein Lyriker im Zeitalter des Hochkapitalismus* (Frankfurt am Main: Suhrkamp, 1969), p. 189; rendered in English in Bernd Witte, "Benjamin and Lukács: Historical Notes on the Relationship Between Their Political and Aesthetic Theories," *New German Critique* 5 (Spring 1975), 10–11.

discoveries in science and technology encounter, then it must be recalled that every fascist 'order' begins by wiping out the individual."[8] Wolf's defense of individualism should not be misunderstood, however, as a depoliticization of art. On the contrary, it is in response to the over-politicization of culture, and to the synchronization of substructure and superstructure, that Wolf insists on radicalizing culture so that it becomes labor itself rather than its legitimation. She calls for a form of writing which "has the courage to think of itself as an instrument—sharp, accurate, attacking, changing, to be used as a means, not as an end in itself."[9] Inherent in her critique of the literary heritage of the Englightenment is also the insight into the ultimate reification of Enlightened Man. Referring implicitly to Galileo's faith in reason and science, she writes: "The unrivalled optimism of this mechanism, not to be doubted even by those it crushes, is obvious: I perish, but it remains and moves."[10]

Striking in Wolf is an almost Adorno-like distrust of media and the culture industry. Both reading and writing, she insists, are highly individual processes; prose should try to be "unfilmable,"[11] and thus, it signals a utopian protest against the status quo. This reserve toward the media is shared by many avant-garde writers in the GDR and can be explained by the peculiar nature of literary production and reception there. In contrast to the polarization of high and low culture in Western capitalist societies, GDR society is dominated by what can be termed a "median culture," and, political constraints notwithstanding, the best writers are often also the most popular. This is the case with Christa Wolf, whose books tend to be sold out within the first week and who, despite the opacity of her narrative style, is widely read among workers and youth. Given the history of cultural policy and the conviction of writers that literature be politically current as well as entertaining, literature in the GDR exhibits a journalistic quality, in the good as well as the bad sense of the word, and it has assumed much of the informational and entertainment value that in our society is limited to newspapers, magazines, and television. Indeed, the average GDR citizen, when not watching West German TV, is much more likely to be politically motivated by reading a current novel than by opening the sterile pages of the newspaper or flicking the channel on TV. The same sterility holds true, for the most part, for theater and film. Small wonder that Christa Wolf sees in the private experience of literature, where the decision to read is a relatively free and spontaneous one in contrast to the theater

[8] Christa Wolf, *The Reader and the Writer,* trans. Joan Becker (New York: International Publishing Co., 1978), p. 211.

[9] *Ibid,* p. 201.

[10] *Ibid,* p. 195.

[11] *Ibid,* p. 193.

visits organized by school or work place, something that "entices and encourages people to achieve the impossible."[12]

A similar tendency toward de-institutionalizing culture is represented by the dramatist Heiner Müller who has turned from the didactic Brechtian tradition to the model of the learning play. Müller's insistence on subverting the dominant forms of mass culture has led him into relative isolation, however, for there is no theater or public sphere which can accommodate him. At least for now, Müller represents the negative implications of Wolf's demand for a radical literature that "entices . . . to achieve the impossible." Like Wolf, though in a more systematic fashion, Müller focuses on the problem of Enlightenment instrumental rationality mediated through the Second International and Leninism. Where Wolf comes to her critique of this tradition through Ernst Bloch and thus stresses the constructive potential of fantasy and subjectivity, Müller is informed by Adorno, Horkheimer, and Benjamin. He relies on the principle of a negative dialectic, which, in Benjamin's words, is to blast open the continuum of history and thus offers "a revolutionary chance in the fight for a repressed past."[13] Where his predecessor Brecht addressed himself to historical event, Müller treats history as ideology—and thereby seeks to subvert it. His adaptation of Aeschylus' *Prometheus Bound* appeared in 1969, one year after the Soviet invasion of Czechoslovakia. In this historical context Prometheus lends himself ideally as Marx's foremost "saint and martyr" who "recognizes man's self-consciousness as the highest divinity." Indeed, it is this attempt to rescue the Promethean concept of justified rebellion that underlies the works of most writers responding critically to the events of 1968.

Müller, however, inverts this approach and lays bare the problematic of the Prometheus myth itself. Hardly deviating from Aeschylus' model, he uncovers in Prometheus "the contradiction between productivity and conceit, consciousness and suffering, immortality and the fear of death."[14] Prometheus is inconceivable without the gods who rule over and within him, and as such he still affirms the referential framework which he seeks to reject. In this sense he is as much a hindrance to progress as he is its vehicle. Horkheimer and Adorno write in *Dialectic of Enlightenment:* "For the Enlightenment, whatever does not conform to the rule of computation and utility is suspect. . . . Every spiritual resistance it encounters serves merely to increase its strength. Which means that enlightenment still recognizes itself even in myths. Whatever myths the resistance may

[12] *Ibid*, p. 212.

[13] Walter Benjamin, "Theses on the Philosophy of History," in his *Illuminations*, trans. Harry Zohn, ed. Hannah Arendt (New York: Harcourt, Brace & World, 1968), p. 263.

[14] Heiner Müller, Note to *Prometheus*, in his *Geschichten aus der Produktion*, 2 (Berlin-West: Rotbuch Verlag, 1974), p. 55.

appeal to, by virtue of the very fact that they become arguments in the process of opposition, they acknowledge the principle of dissolvent rationality for which they reproach the Enlightenment. Enlightenment is totalitarian."[15] Where a critic like Wolf Biermann makes clear distinctions between forms of reaction and revolt, Müller finds the deformations of the past within progress itself. Instead of suggesting like Biermann: "I'm the individual / the collective has isolated / itself from me,"[16] he places himself within the text as historical product as well as critic. In his reflections on the events of 1968 implicit in his *Prometheus* adaptation, he identifies as much with the gods as he does with the man who despises them. And indeed, without the power and knowledge he shared with the gods, Prometheus could never have become a rebel.

The link between the Prometheus myth and the bourgeois Enlightenment is more explicit in Müller's play *Cement,* an adaptation of Feyodor Gladkov's novel (*Cement,* 1925), which he wrote in 1972. Here the liberation of Prometheus by Herakles is one of three mythological passages in the play which serve to both extend and complicate the dramatic action. *Cement* treats the reconstruction period in the Soviet Union after the Civil War where, after Lenin's prognosis, technology and the revolutionary proletariat were to walk side by side. The problematic inherent in this concept is exposed by Müller not only as a critique of the Soviet reconstruction, but also as being inherent in the nature of revolutionary transformation itself, which must absorb the knowledge of a past era in the creation of the new.

Liberation of Prometheus

> Prometheus brought lightning to humans, but did not teach them to use it against the gods, because he shared his meals with the gods, and they would have been less sumptuous if shared with humans. Either because of his deed or because of his omission, the gods ordered Prometheus to be fastened by Hephaestus the smith to the Caucasus, and a dog-headed eagle ate daily from his liver, which always grew back. The eagle considered him to be a partly edible rock formation, capable of small movements and, especially when being eaten, of discordant song, and emptied its bowels on him. Its feces were his food. He transformed them into his own feces and passed them on the stone beneath him. Thus three thousand years later, when Herakles his liberator reached the top of the uninhabited moutains, he was able, even from a great distance, to distinguish the prisoner clearly, shining white with bird feces. But he was repelled again and again by the wall of stench, and he circled around the massif another three thousand years. Meanwhile the dog-

[15]Max Horkheimer and Theodor W. Adorno, *Dialectic of Enlightenment,* trans. John Cumming (New York: The Seabury Press, 1975), p. 6.

[16]Wolf Biermann, "Ruthless Cursing," trans. Jack Zipes, *New German Critique* 10 (Winter 1977), 7.

headed eagle continued to eat the prisoner's liver and to nourish him with his feces, so that the stench increased to the same degree that the liberator became accustomed to it. Finally, favored by a rain that lasted five hundred years, Herakles was able to approach within shooting range. He held his nose shut with his hand. Three times he missed the eagle, for when he took his hand away from his nose to stretch his bow he closed his eyes involuntarily, stupefied by the wave of stench that assaulted him. The third arrow wounded the prisoner lightly on the left foot. The fourth killed the eagle. Prometheus, it is told, wept loudly for the eagle, his only companion for three thousand years and his source of food for twice three thousand. Shall I eat your arrows, he screamed, and, forgetting that he had once known other nourishment: can you fly, peasant, with your feet of manure. Then he vomited from the stable smell which had clung to Herakles since the time he had cleaned the Augean Stables, because the manure stank to the heavens. Eat the eagle, said Herakles. Prometheus could not grasp the meaning of his words. He knew that the eagle had been his last link to the gods, its daily pecking his remembrance of them. More flexible than ever in his chains, he cursed his liberator, called him a murderer and tried to spit in his face. In the meantime, Herakles, bent over with nausea, looked for the fetters which bound the raging Prometheus to his prison. Time, weather, and bird feces had made flesh indistinguishable from metal, and both indistinguishable from the stone. Now loosened by the more violent movements of the prisoner, the fetters became discernible. They had been eaten by rust. Only on his sex had the chain and flesh grown together because, at least during his first two thousand years on the rock, Prometheus had masturbated occasionally. Later he must have forgotten even his sex. The liberation left a scar behind. As unarmed and exhausted from the millennia as he was, Prometheus still could have freed himself easily had he not feared the eagle. His behavior during the liberation demonstrates that he feared freedom more than the bird. Roaring and spitting with rage, he protected his chains from the grip of the liberator. Once freed, he howled on his hands and knees from the torment of trying to crawl with his numb limbs. He screamed for his quiet place on the rock beneath the wings of the eagle, where nothing moved unless shaken by an occasional earthquake decreed by the gods. Even after he was able to walk upright again, he struggled against the descent like an actor who does not want to leave his stage. Herakles had to carry him down from the mountains on his shoulders. The descent to the populated regions lasted another three thousand years. The gods rooted up the mountains, so that the descent through the swirling fragments of stone was more like a plunge. Herakles carried his precious booty snuggled like a baby against his chest in order that he not be harmed. Clinging to his liberator's neck, Prometheus indicated the directions of the shots, so that they were able to dodge most of them. Meanwhile, screaming loudly to the heavens which were darkened by the swirl of stones, he declared his innocence in the liberation. The suicide of the gods followed. One after the other they hurled themselves down from the heavens onto Herakles' back, and shattered in the rubble. Prometheus worked his way back onto the shoulders of his liberator and assumed the pose of the victor who rides in on a sweating horse to meet the cheers of the populace.[17]

Significantly, Müller equates Prometheus with Enlightenment and bourgeois revolution—the technocrat who refrained from giving the masses control over their labor because he still wished to mingle with the ruling gods. It is the unidealistic and unwilling Herakles who is thus fated to carry out the onerous job of liberating the former rebel and his knowledge, and on whose shoulders Prometheus is unbound—and yet bound. In contrast to Marx, who came out of the Enlightenment and from whom Prometheus was both *homo faber* and rebel, Müller separates the two concepts and uncovers their individual historical problematic. Thus he demonstrates that, like the bourgeois Enlightenment itself, these concepts have been transported into the twentieth century and abstracted as ideology by the superstructure. Nonetheless, as evidenced by the dialectic between the sweating liberator Herakles and Prometheus on his back posing as victor, this is both an oppressive and a necessary superstructure within the historical continuum. In this respect Müller acknowledges the historical validity of the Enlightenment tradition while at the same time critiquing it. Whereas for Horkheimer and Adorno the Enlightenment dominated all modern history, Müller lays bare its own historicity: in the new myth of Herakles, namely in the *forces* of labor, he finds a vehicle whereby Enlightenment and ideology are no longer locked in the abstract impasse of a negative dialectic.

Nevertheless, the new myth of the liberator Herakles, which suggests the vanguard of the proletariat as defined by Leninism, is inconceivable without the myth of Prometheus literally on its back. In this sense Müller exposes the alienation within Leninism itself, whose labor is as bound by the weight of the past as it is yet separated from the people as a whole whom it seeks to serve. It is this alienation that, not only for Heiner Müller but for all neo-Marxist writing in the GDR, represents the dilemma of Marxist praxis since Marx. Given the separation of production and revolt, which for Marx were still unified, the moment of revolutionary transformation is no longer conceived by these writers in terms of Marxist categories of production, but tends to be articulated as subjective existential experience. Recalling Walter Benjamin's notion of blasting open the historical continuum, a recent text by the young writer Thomas Brasch summarizes in parabolic form this attempt to salvage the underlying Marxist concept of radical change by exploding its mediated historical framework. The worker Fastnacht, caught up in the gears of

[17]Heiner Müller, *Zement,* in his *Geschichten aus der Produktion,* 2, pp. 84–86. English translation of *Cement* by Helen Fehervary, Sue-Ellen Case, and Marc D. Silberman, *New German Critique,* Supplement to Issue No. 16 (Winter 1979), 25–27.

productivism and bureaucracy, is driven to the only place where he can escape and make history for himself—suicide in the washroom, a literary *topos* extraneous to the categorical landscape of Marxism:

> Bolt the door, up on the toilet seat, belt out of the pants, tie it to the flushing chain, leather around the neck. Where are you going, said Marxengels. What's going to happen to the German working class if you hang yourself here in the washroom? What's going to happen to me if I don't hang myself, said Fastnacht. After capitalism, said Marxengels, comes communism or advanced technological barbarism. Get down, Fastnacht. Get off my back, you bearded prophet, said Fastnacht. This place is occupied.[18]

Exposing not only the Promethean myth of the Marxist tradition but the mythical character assumed by the tradition itself, Brasch goes one step further than Müller and brings the dialectic of myth to a final impasse. In sublating the myth, however, he does less to separate himself from the essential Marxist notion of praxis than to reassert it: production and revolt are united once more, not in the ideology of Enlightenment but in the concrete experience of the individual self-determining human activity and change. The metaphor of self-negation suggests a radical break with the accumulation of culture, and simultaneously the freeing of a utopian potential which has been repressed by the dead weight of the past. This attempt at radicalizing culture, however, can only assert itself as a negative dialectic in the political-economic reality of the GDR today. Where the legitimation of the Marxist tradition increasingly takes the forms of scientific objectification, technocracy, and the bureaucratic paper tiger, any viable self-critique is relegated to the ideological periphery of Marxism and reduced politically to the level of dissidence. And so it has been for Brasch whose insistence on the crisis of Marxism in the GDR ultimately forced him into exile.[19] Whatever the historical consequences of a situation like Brasch's which exemplifies the impasse of cultural critique in the GDR, the fact remains that a unity of production and revolt can no longer be articulated in the categorical language of a system that suspends the very dialectic in which its myth and tradition rest.

[18]Thomas Brasch, "Fastnacht," in his *Vor den Vätern sterben die Söhne* (Berlin-West: Rotbuch Verlag, 1977), p. 100. My translation.

[19]For more detailed accounts of Brasch's history, see *New German Critique* 12 (Fall 1977): Helen Fehervary, "Thomas Brasch: A Storyteller after Kafka," 125–32; and "Interview with Thomas Brasch," 141–68.

Affirmative and Negative Culture: Technology and the Left Avant-Garde*

DAVID BATHRICK

The artistic avant-garde has always had little respect for entrenched authority, even when that authority claims for itself revolutionary intention. Its history within the traditions and institutions of socialist societies constitutes a legacy of mutual distrust, sometimes open warfare, and repeated repression—but also a return of the repressed. Indeed, this continued dialectic of repression and revolt is precisely what distinguishes the avant-garde in Eastern Europe from its function in advanced Western capitalist societies, where, as most critics agree,[1] the integrative mechanisms of the culture industry have long since relegated the historical avant-garde to just that—the backwaters of cultural history.

The hostility within Marxism and among socialists toward what has variously been called naturalism, modernism, decadence, formalism, or—most pejoratively—"avantgardism" finds its earliest antecedents in nineteenth-century social democracy[2] and its more programmatic begin-

*This essay first appeared in *Social Research*, 47: 1 (Spring 1980), 166–187.

[1]Critics in both Europe (Hans Magnus Enzensberger) and the United States (Leslie Fiedler, Irving Howe) have proclaimed the "death of the avant-garde"; their view has recently been elaborated into a "theory of the avant-garde" by the West German scholar Peter Bürger in his *Theorie der Avantgarde* (Frankfurt am Main: Suhrkamp, 1974).

[2]Franz Mehring, the socialist historian, was particularly outspoken and influential in his attack upon such modernist movements as naturalism and equally adamant in his defense of Schillerian drama as the model for revolutionary proletarian art.

nings in the aftermath of the October Revolution. If Lenin and Trotsky differed in their individual assessments of movements such as Futurism or artists such as Vladimir Mayakovsky, they were one in their emphasis upon classical bourgeois culture as the cornerstone for any socialist culture. In the words of Trotsky: "The working class does not know the old literature, it still has to commune with it, it still has to master Pushkin, to absorb him, and so overcome him. . . ."[3] Lenin was even more adamant in opposing the avant-garde *Proletkult* movement and calling for literacy as the first cultural priority of the revolution:

> While we have been palavering about proletarian culture and its relation to bourgeois culture, the facts indicate that not even bourgeois culture is faring very well with us today. . . . Proletarian culture simply has to be the systematic further development of the sum of knowledge which mankind has accrued under the yoke of capitalist, feudal and bureaucratic social orders.[4]

Of course, Lenin's nod to Tolstoy and the classics stemmed as much from the immediate exigencies of the postwar period as it did from any general theory of culture or even from personal taste. The process of industrialization lent priority to mass education in a society in which 70 percent of the population were illiterate. But there were important political reasons as well. The activities of highly vocal left groupings around Bogdanov's *Proletkult,* Mayakovsky and the left Futurists in LEF (Left Front of Literature) and the Constructivists constituted a potentially independent and hence politically threatening element within the precarious social order. Their call for self-organization and autonomy as the basis for creating any truly revolutionary culture clearly ran counter to the move toward centralization and party control by the Bolsheviks. Seen in this light, the weaving of bourgeois tradition into the fabric of an emerging cultural policy was very much a part of Lenin's struggle in the 1920s to establish political hegemony. As such it provides an early example of traditionalist aesthetics being used as a form of political legitimation—a policy which was to reach its more elaborated codification with the entombment of the nineteenth-century novel in the prescriptive edicts of Socialist Realism in 1934. Thus the bourgeois realists Tolstoy, Balzac, and Thomas Mann, not Franz Kafka or James Joyce or even Bertolt Brecht, were to provide the models for literary policy as it developed from the first five-year plan (1928–33) into the postwar period; or as it was articulated in the literary theories of an Andrey Zhdanov or a Georg Lukács or a Johannes R. Becher. Since they viewed modernism as

[3]Leon Trotsky, *Literature and Revolution* (Ann Arbor: The University of Michigan Press, 1960), p. 130.

[4]V. I. Lenin, "Pages from a Diary" (p. 692) and "The Tasks of the Youth Leagues" (p. 413) in *Selected Works* (Moscow: Progressive Publishers, 1964), vol. 3.

decadent and destructive of culture itself, Marxist theory and socialist practice have often been said to share much with conservative critics of the bourgeois mold. Both the Marxists and certain reactionary defenders of high culture, it is argued, view in the assault upon aesthetic autonomy and the diffusion of the organic work of art modernity's (capitalism's) threat to the very foundations of enlightenment culture as the realization of reason and harmony. In opposition to such an approach, this essay seeks to explore the dynamics of cultural struggle within the specific historical framework of a developing socialist cultural policy. While it is indeed true that avatars of philosophical idealism or biological positivism have helped shape the key schools of Marxist aesthetics (Lukács and Plekhanov), it is precisely the *function* of such ideas which would allow us to constitute their ultimate meaning.

From Socialist Realism to Post-Brechtian Theater

The German Democratic Republic (GDR) may serve as a case in point. The cultural policy directives regarding "formalism" developed under Ulbricht in the early 1950s obviously inherited a political framework and institutional biases stemming directly from the Soviet experiences of the 1920s and 1930s. In one important way the similarity between the goals of the Bolsheviks and those of what has been called the historical avant-garde marks their relationship as problematic. From its very inception in the October Revolution, Lenin's cultural policy, while politically at odds with the avant-garde, claimed that it was coopting and indeed realizing, *not* repudiating, the revolutionary project of such artistic movements as Futurism, Dadaism, and Surrealism. For instance, if the advocate of production aesthetics Boris Arvatov called for the "complete melding of the artistic forms of everyday life; the complete dipping of art into life,"[5] to be central to to a new revolutionary art, then the joining of bourgeois spirit to a proletarian body in Lenin's policy *also* claimed to be *Aufhebung* (sublation) of the values and ideals of affirmative culture through their absorption and further realization in everyday socialist reality. In this regard, even Walter Ulbricht's claim that the GDR was "Faust Part III"[6] is but a somewhat far-fetched version of that strain of the artistic avant-garde which has always sought to transcend the dichotomy of art and life. To be sure, Ulbricht's and, let us say, Sergey Tretyakov's notions of *Aufhebung* here are themselves somewhat dichotomous. Where the

[5]Boris Arvatov, *Kunst und Produktion—Entwurf einer proletarisch-avantgardistischen Ästhetik (1921–1930)*, edited by Hans Günther and Karla Hielscher (Munich: Carl Hanser, 1972), p. 27.

[6]In a speech before the National Council of the National Front on March 25, 1962, reprinted in *Neues Deutschland*, March 28, 1962.

Russian theorist was iconoclastic in his assault upon the institution of traditional art, which he would use as "compost"[7] for the new order, the cultural policy of the SED sought to terminate bourgeois autonomous culture and simultaneously to reinstitute it in what Johannes R. Becher has called *die Literaturgesellschaft* (the literary society).

The first important GDR document dealing with the avant-garde was a declaration by the Central Committee of the SED on March 17, 1951 entitled "The Struggle against Formalism in Art and Literature, for a Progressive German Culture." The categories it employs clearly reflect its historical origins. Formalism or "cosmopolitanism," we are told, is a form of "American cultural barbarism" which, when practiced in the new society, leads to a "rupture with art itself, a destruction of national consciousness and an indirect support to the war policies of world imperialism."[8] What is important here is not the content but rather the *structure* of a political relationship. The rhetoric and posture in such phrases as cultural barbarism and cosmopolitanism were clearly a part of the then-prevailing Cold War climate and were soon to disappear with the beginning of destalinization and a greater focus on political conflicts within the social order. What was not to disappear was this re-establishment within GDR socialism of the classical antagonism between the institution of affirmative culture on the one hand and its potential subversion at the hands of some form of modernism—an antagonism transferred now into the legitimacy struggles of "a state that is not supposed to be."[9] The repoliticizing of culture here is of course a two-edged sword. While the welding of "national form to a socialist content" (Ulbricht) clearly served the nationalist identity claims of the new order, it also meant that an iconoclastic alternative to that form was perforce a challenge to the larger political order. Nobody understood this dialectics of legitimation[10] better than Bertolt Brecht, who when summoned to the highest ministers of state to discuss the "formalist" and ideological deviations of his and Paul Dessau's opera *Lukullus,* asked with characteristic *double entendre:* "Where else in the world can you find a government that shows such interest in, and pays such attention to, its artists?"[11]

If the framework of official cultural policy institutionalized the

[7]Sergey Tretyakov, *Die Arbeit des Schriftstellers* (Reinbek: Rowohlt, 1972), p. 22.

[8]"Der Kampf gegen den Formalismus in Kunst und Literatur, für eine fortschrittliche deutsche Kultur," in *Dokumente der SED* (Berlin, 1952), 3, 431–446.

[9]Subtitle of an early widely read book on the GDR by Ernst Richert, *Das Zweite Deutschland—Eine Staat, der nicht sein darf* (Gütersloh: Sigbert Mohn, 1964).

[10]See David Bathrick, "The Dialectics of Legitimation: Brecht in the GDR," *New German Critique* 2 (Spring 1974), 90–103.

[11]Quoted in Jürgen Rühle, *Das gefesselte Theater* (Cologne/Berlin: Kiepenheuer & Witsch, 1957), p. 243.

conflict between two bourgeois traditions, high culture and the avant-garde, the realization of that conflict has varied with changing historical realities in the GDR. Within the Cold War of the early 1950s, it was Brechtian "formalism" (epic theater generally, productions of *The Mother*, *Lukullus*, and *Mother Courage*) and Hans Eisler's irreverent treatment of *the* classic in his adaptation of *Faust* which were seen indirectly to undermine the state as bearer of a humanist-democratic tradition. The cultural revolution of the post-Stalin 1950s and early 1960s saw the playwrights Peter Hacks, Heiner Müller, and later Volker Braun draw on the constructivist traditions of the Soviet Union and the political-cultural avant-garde of Weimar (Brecht, Walter Benjamin, Berta Lask, Erwin Piscator) to develop and agitprop drama critical of the socioeconomic failures of the "construction years." Finally the emergence of the alienation problematic in the Kafka debate of 1963 was accompanied in the GDR by "subjectivist" experimentation in prose (Christa Wolf's *The Divided Heaven* [1963] and later *The Quest for Christa T* [1968]) and the poetic activities of such "modernist" poets as Karl Mickel, Günter Kunert, and Sarah and Rainer Kirsch. Epic theater, *Proletkult*, Joycean stream of consciousness, lyrical experimentalism—in each case the struggle has been political in content and has resulted in temporary restrictions. But it has also ultimately led to the acceptance of this or that model of the avant-garde as compatible with what now appears to be an ever-increasing elasticity in the notions of "cultural heritage" and socialist realism. Thus it might have come as no surprise when, as part of the VIII Party Congress in 1971, Secretary General Erich Honnecker seemed to culminate this process with the statement, "In the area of art and literature there are no longer any tabus!"[12] For many, this was merely the final recognition that the avant-garde had been neutralized the way it would always be neutralized—not through repression but through repressive tolerance. The question we must answer now is whether this is true and, if not, what are the parameters around which the process has continued into the 1970s.

Certainly the expulsion of Wolf Biermann in November 1976 and the subsequent exiling and silencing of a number of leading poets and writers was a clear indication that "tabus" still exist and will be enforced. But neither Biermann nor Thomas Brasch nor Jürgen Fuchs nor even Jurek Becker ran into trouble primarily because of modernist tendencies in their writings. Thus, let us reformulate the question differently: Does the avant-garde in any of its historically recognized forms and roles still function politically in the GDR and, if so, what does it look like and what are its premises? I should like to argue that a form of the avant-garde does still exist in the GDR and that is premises and potentialities may best

[12]Reprinted in *Neues Deutschland*, Dec. 18, 1971.

be examined in the works and function of its leading playwright, Heiner Müller.

In Müller we have a writer who comes from and was schooled in the Brecht tradition but who, in a number of fundamental ways, has inverted and questioned that tradition as a viable mode of revolutionary expression. Born in Saxony in 1929, Müller was the first of the postwar GDR playwrights to confront directly and brutally the historical realities through which he had lived. Certainly the major phases of his extraordinarily prolific career have at once chronicled and helped shape pivotal developments in GDR drama itself. The early production plays (*The Wage Shark,* 1956; *The Correction,* 1959; *The Construction,* 1961) appeared within the political turmoil of the "construction" years and represent an aesthetic and political break with the mythologies of Socialist Realism dominating literature at that time. When in the 1960s controversial writers were discouraged from treating contemporary problems, Müller turned to classical adaptations (*Philoctetes, Herakles 5, Oedipus Tyrann, Prometheus, Macbeth*) which *mutatis mutandis,* turned out to be no less political or controversial than some of his earlier plays. It was also at this time that he began to write his history dramas, works which attempt to deal with the legacies still present in the GDR, be they from the German (*Germania, The Battle, Construction, The Peasants*) or Russian traditions (*Cement*). In scope of vision, political orientation, and innovative technique Müller has clearly established himself as the heir apparent to Bertolt Brecht, and his being commissioned by Brecht's Berliner Ensemble to write a play for the 1972 season would seem to have sealed that claim.

Yet despite or perhaps because of his debt to the maverick Brecht, Müller has become increasingly skeptical over the years concerning the relevance of the Brechtian tradition for a critical theater in the socialist world. The reasons for this disenchantment may be traced both to Brecht himself and his reception in the GDR. At the level of cultural policy, it is clear that with the canonization of Brecht in the museum-like Berliner Ensemble and with the resolution of the now famous Brecht-Lukács (modernism-traditionalism) debate[13] in at least a stand-off, Brecht and the tradition he represents (montage, estrangement technique, epic theater) have been integrated into the theoretical and practicing life of socialist art. But the problem for Müller may also be located within Brecht's own writing and thought. If the early proletkultist Heiner Müller looked to Brecht's epic theater and objectivised notions of historical progress as the basis for his dramaturgy, Müller's later work increasingly

[13]For a discussion in English of the Brecht-Lukács debate, see Eugene Lunn, "Marxism and Art in the Era of Stalin and Hitler: A Comparison of Brecht and Lukács," *New German Critique* 3 (Fall 1974), 12–44.

questions such a "macrostructural" aesthetic for what it represses in terms of historical and personal experience:

> Now we must concern ourselves with microstructures. For that Brecht offered forms and techniques only in his early works . . . but not in the classical plays. That is why they are so sacrosanct and boring.[14]

Sacrosanct, boring, Müller says, *classical*—epithets which the avant-garde has always hurled at the establishment. And Müller's critique of Brecht often has the attributes of such a clash. But again we must ask, why Brecht? Are they both not Marxists of the avant-garde who have inverted in similar ways the precepts of orthodoxy to develop a poetics of consciousness? To answer this question we must look beyond Brecht the individual to the artistic and political tradition he represents.

Avant-Garde Aesthetics vs. Leninist Aesthetics

While Brecht's writings and theatrical activity as a whole represent the most coherent "materialist" alternative to the official aesthetics which was to emerge from the avant-garde within the socialist tradition, it is also true that much of the impetus and many of the practical models for not only Brecht but also a large number of left intellectuals of the Weimar period were to come from the Soviet avant-garde of 1917–23. In this regard, the theoretical achievements of Walter Benjamin and Bertolt Brecht throughout the later 1920s and 1930s constitute an important link between the earlier Soviet experiences and the reemergence of that tradition in the GDR of the 1950s and in West Germany a decade later.

Central to the whole notion of aesthetics to spring from the Russian postrevolutionary avant-garde was an integral link between cultural and social production. While the Formalist and Futurist movements and many of the leading poets (Mayakovsky, Burliuk, Khlebnikov) were very much in evidence in the early 1900s, it was the social, political, and economic energies released by the revolution which helped reorganize their potential and whole orientation. For instance, where the linguist experimentalists around Viktor Shklovsky and the OPOYAZ group once sought to close the gap between life and art by subverting traditional aesthetics at the level of syntax and image in order to "liberate" letters and words from the "hegemony" of context, their left Futurist progeny aimed their attack at the very act of representation itself. Basic to the theories developed most coherently in the writings of Tretyakov and Arvatov was a notion of art, not as something autonomous and self-contained, but rather as the "organization of everyday life," as an appropriation of nature—as a form of material production. Indeed, by recasting artistic

[14] *Die Zeit*, March 17, 1978.

creativity within the categories of labor power and material production, the LEF theorists strove for a materialist aesthetic which challenged the premises of bourgeois, but also implicitly of Leninist, aesthetics. What links bourgeois and Leninist notions of consciousness and art is an emphasis upon an ontology of knowledge, upon consciousness as a secondary, reflected, and passive repository for finished "products" of thought. The Futurists questioned this by seeking to break with what Arvatov pejoratively referred to as "easel art" (art as supplementation of a disharmonized, i.e., unorganized reality) and by establishing a threefold relationship to the forces of production at large: (1) as an activity which appropriated the physical materials and organizational principles of the industrialization of everyday life—film, radio, the other mechanical media; (2) as an activity which functioned within and dissolved the the qualitative differences between the productions of everyday life—the production of useful objects, the design of objects and cities, the aesthetics of functionalism, etc.; (3) as an activity which availed itself of the *formative* principles of industrial and technological modes of production—montage, etc.[15]

Metaphors of production and the importance of science and technology for art were of course not original to the Russian avant-garde. The Italian Futurists also saw in the machine both an irreverence toward the sensibilities of traditional aesthetics and a thing of beauty in itself, and their songs of praise to technical weaponry clearly anticipate the further implementation of such an "aestheticization of politics" (Benjamin) as it developed in Italian and German fascism. Even the Surrealists grasped in the concept of production an analogue to the furious and machine-like workings of the poetic unconscious.

What distinguishes the Russsians from either the Italian Futurists or the French Surrealists is their attempt to link artistic activity directly to material societal production and thereby the activities of the artist to a whole set of "laws," processes and energies perceived to be the inexorable, natural movement of history itself. This latter expansion of the concept of production is important for Bertolt Brecht as well, who, like Arvatov and others, conceived of production aesthetics both in regard to the techniques employed and in art's relationship to the unfolding of social production as a whole. He referred to socialism as "the grand production," by which he meant a social system which would release the *entire* creative and working potentials of human beings, thus dissolving the distinctions in labor processes.[16] It is also clear that his estrangement effect, while initially borrowed from and similar to Shklovsky's "making

[15]Arvatov, *Kunst und Produktion*, pp. 52–64.

[16]For a thorough if somewhat uncritical discussion of production aesthetics and its relationship to Brecht, see Heinz Brüggemann, "Aspekte einer marxistischen Produktions-

strange," was developed and expanded within Brecht's later range of thought into a category of mental production. As a "technique of thinking" it worked upon the seemingly natural and given world of fetishized objects and appearances in very much the way tools and machines are employed "critically" in the appropriation of the natural world. It, too, then was a force of production, a part of the "ever proceeding emancipation of humans from nature on the ground of nature itself." Thus while Brecht, in contrast to the LEF theorists, did not envision the dissolving of art and life into mere activity, and while he always emphasized centrally that art remained something constructed and hence second nature, his holding to and restatement of the theory of aesthetic production carried many of the political implications of the Soviet avant-garde into the landscape of the GDR.

And what were the political implications of this aesthetic? Why was *Proletkult,* or, as it was called in the 1950s, "Neorappism,"[17] to remain a thoroughly negative epithet within cultural policy? First, and most important, the anchoring of artistic production within the base as a coforce of production meant a repudiation of the base-superstructure model and with that the political infrastructure fed by Socialist Realism. The concept of reflection is important for Marxist-Leninism not merely because in its seeming duplication of reality it claims to provide a more communicable ("realistic," "popular") representation of the world, but because the representation itself is a replica of pre-established modes of being and knowing. In two interrelated ways works of Socialist Realism are the results of *prefabrication:* As works reflecting "reality" they present a cognitive organization after the fact, that is, one that conforms and is subordinate to a set of "objective," "natural" processes. Second, the view of these processes is itself to be informed by an objective science which "knows" already certain outcomes by virtue of the categories it employs (inevitable coming of socialism from the inevitable collapse of capitalism, etc.). Once given these two interfacing premises—the inexorableness of "socialist" history and the infallibility of the science which will know it—it follows that the party installs itself as the omniscient mediator of the whole process.

Brecht's aesthetic challenges this in a number of important ways. The making of the author into a producer (Benjamin) shifts the whole emphasis of a literary epistemology from a reflected "content of knowl-

---

Ästhetik," in *Literaturwissenschaft und Sozialwissenschaften, 4,* edited by Heinz Schlaffer (Stuttgart: Metzler, 1974), pp. 109–144.

[17]RAPP was the Russian Association of Proletarian Writers at the end of the 1920s and beginning of the 1930s. The reference to RAPP at this later stage is clearly using it as generic for all the avant-garde groupings from the Proletkultists through LEF, regardless of their political differences on any number of issues.

edge" to an *active* "critical" reorganization of experience. This in turn affects all the categories that Brecht employs. On the one hand, by grounding literary production in social production as a whole he clearly retains the Marxian notion of the unfolding of the forces of production through the process of appropriating nature (*techné*) as basic to a concept of historical progress. At the same time, by making the author a producer and instituting a *direct* two-way relationship between an "epic" voice of reason (producer I) and the critically reasoning audience (producer II), Brecht broadens his notion of production to include the production of meaning and consciousness as well. It is this more democratic emphasis upon the shared production of meaning—basic to both the concept of estrangement and the principles of the *Lehrstück* (learning play)—which implicitly challenges the hierarchy of the *one* science (Brecht speaks of his audiences as the "children of the scientific age") and which links him to the "anarchy" of the leftist avant-garde.

Looking at the theater of the GDR in the 1950s and 1960s, we see a whole tradition of drama production which has been influenced by Brecht and which was not appreciated by officialdom for its efforts. Although not all these artists were rooted within the tradition of epic theater or were necessarily adherents of the method of estrangement, each shared the emphasis, indeed euphoria, concerning the emancipatory and democratic potential of an unfettered production and particularly the notion of the writer as a coproducer, not merely a registrant, of the social process. Hartmut Lange (*Senftenberger Stories,* 1959), Peter Hacks (*The Cares and the Power,* 1959), and Volker Braun (*Kipper Bauch,* 1965) all focus upon contradictions within the area of material production as a way of dealing thematically with contemporary problems in the GDR. In each case the choice of a factory as the locus of the action allows the playwright to at once criticize the failings of social organization while at the same time affirming the ultimately progressive and optimistic role of production itself for the building of a better world.

Certainly the most important of these writers was Heiner Müller himself. His first production play, *The Wage Shark,* literally picked up where Bertolt Brecht left off. In the early 1950s, toward the end of his life, Brecht had made plans to write a *Lehrstück* treating the "heroic" deeds of the GDR activist and medal winner, Hans Garbe. A bricklayer in a construction brigade, Garbe gained national attention and ultimately poetic immortality by singlehandedly repairing a damaged oven at the Siemens-Plania plant in East Berlin without having the surrounding ovens shut down and hence without stopping what at that time was a vital segment of GDR production. Surely this was the material for socialist mythmaking, which may have been the reason Brecht dropped the project. In any case, it was Müller who drew on the Garbe material to create a work which has become a classic for the genre. Composed of fifteen short scenes and written in a sparse, abstracted, everyday dialogue, *The Wage Shark* represents an original appropriation of the Brechtian and constructivist traditions in order to look critically at the inadequate conditions of the early building years in the GDR. Significant for Müller is the fact that, unlike the myriad of other Socialist Realist treatments of this and similar subjects, the dynamic of the play is not simply organized around production as a theme or the ontology of character in the form of a Stakhonovite hero who propels the action. Hence in Müller's play, Barka/Garbe is a negative, not a positive hero, a man who had once denounced a group of saboteurs in the Third Reich for disrupting production and whose obsessive and egotistical compulsion to work even now is disruptive to the work relations in the newly founded society. Nor does Müller rely on a linear plot narrative which might move toward a classical synthesis or in any way resolve situations harmoniously. The montage of scenes here stand in discontinuous and negative relationship to one another. The stage directions at the end of Scene 1 present in symbolic pantomime form a struggle *against,* which in its constellation with other negative actions and relationships may stand as typical of the whole:

> On the street there appears a signposter who posts a sign with the text: "SED (Socialist Unity Party)—Party for rebuilding" on a ruined building. When he is gone a young man comes along, stands in front of the sign, looks around, rips it down and walks off whistling. Three workers, tired, with satchels under their arms walk across the sign lying on the ground.[18]

Müller is not mirroring (advocating) an attitude for or against the party (or Garbe or the workers against Garbe, etc.). Rather it is the very interaction of negative energies which becomes the operative principle both for the dramaturgy of the play and for the process of industrial production in the larger society. Out of the contradictions at the heart of the industrial process, Müller seems to be saying, is born the new society (and the new aesthetic!).

The Dialectics of the Avant-Garde

In our discussion of the Soviet avant-garde's aesthetics of production and its continuation in the theories and practice of Bertolt Brecht in the 1930s and Heiner Müller in the GDR, we have stressed the ways that such an aesthetic subverts the philosophical and political precepts of a Socialist Realism based on the values of high culture. While such an emphasis explains well how "self-organization" as the thematic and operative

[18]Heiner Müller, *Geschichten aus der Produktion I* (West Berlin: Rotbuch, 1975), p. 16.

principle of art would prove incompatible with the structures of a party policy devoted to mediating all cultural meanings within the social order, it does not help us understand the potential for that same aesthetic to be integrated into the principles of Marxist-Leninism. This latter concern is particularly acute for our discussion of Heiner Müller in the GDR, since his break with Brecht represents in a larger sense a coming to terms with the political realities imbedded in the aesthetics of production. For as much as the productivist avant-garde disrupts and transforms at the level of presentation and re-presentation, it is its historically developed philosophical implications concerning history and subjectivity which have relinked it to an apologetics for the existing socialist order. Basic to its orientation is an unquestioned acceptance of the primacy of humans over nature—the interaction between subject (producer, artist) and object (nature, material for representation)—as *eo ipso* the realization of an enlightened, progressive history. This binding of a theory of production with a theory of history is also of course the theoretical cornerstone of Soviet Marxism for the legitimation of existing socialist societies. If Marxism as a revolutionary social theory stressed the importance of theory (thought, consciousness, art, etc.) as the *critical* self-consciousness of an actual revolutionary process, its transformation into an objective social theory of production grounded in the categories of natural science meant a recasting of the subjective, history-making element into a subcomponent of a naturalized historical process. Understanding the primacy of production for such an ontologized dialectic of history is the key to grasping the paradoxical role of both the left opposition in the 1920s as well as the avant-garde of the same and later periods. In all instances an emphasis upon production was seen as an emancipatory break with scholasticized, sclerotic Marxism.

As much as the Proletkultists and the Futurists challenged the precepts of a dialectic which had become sedimented in a one-way transmission of truth, their unproblematic glorification of production must be seen as ultimately contiguous with the later instrumentalizing of culture and history embodied in Stalin's famous designation of writers and artists as "engineers of the soul." In the Soviet avant-garde we find the same glorification of urbanism and technological culture which was to provide the ideological framework for imposing a modernist, "enlightened," "proletarian" revolution from above upon an agrarian society. Like the mechanists and like Stalin at a later period, the enthusiasts of *Proletkult* voiced uncritical optimism concerning the inherently uplifting and liberatory powers of Taylorism and the Americanization of culture. Thus the fact that Tretyakov, Eisenstein, Mayakovsky, Meyerhold, and others were in opposition to and finally in various ways victims of the emerging Soviet cultural politics of the 1920s should not obscure the fact that their political aesthetic contained as its own internal paradox that very same dialectic of the enlightenment which was the truth of the order they tried to reform.

The Brechtian aesthetic as it was introduced into the GDR contained a similar paradox. On the subversive side, its implicit emphasis upon self-organization was seen and repressed as a challenge to party hegemony in matters cultural. In this regard the critical treatment of Brecht and Benjamin in the GDR is very much in keeping with the repression of any cultural revolutionary program from the Soviet avant-garde through Hans Magnus Enzensberger[19] which calls for self-initiative and pluralism as the constituent parts of a democratized cultural public sphere. Yet although Brecht himself remained to the end an "uncomfortable" Leninist, his embracing of socialist production as the demiurge of historical progress reestablished the "epic" voice of his later plays as the collective subject of an ultimately rational historical process.

It is this "epic voice" and its political implications which Müller begins to question as a viable heritage for radical expression in the GDR. The fundamental difference between the two emerged in Müller's move around 1975 against what he calls the aesthetics of "naturalism": "the expulsion of the author from the text, of the reality of the author from the theater."[20] Brecht is a "naturalist" for Müller because of his relationship to what he writes rather than because of any specific technique or style. While epic theater as an avant-garde, nonrepresentational art form repudiated the simplistic mimetic representations it saw embodied both in nineteenth-century naturalism and twentieth-century Socialist Realism, its sublimation of the authorial self into an "epic" superego derived from the truth and telos of Marxist historiography forfeits a radical subjectivity which Müller seens as essential for a socialist society. For Müller, the Brechtian aesthetic must be seen as the product of an historical period. Brecht's epic persona of the later plays was created within and was an expression of a minority consensus which shared an interest in the destruction of fascism (capitalism). In the course of the 1930s its collective vision was increasingly determined by theoretical premises of "an incomplete Marxism" and propagated the values of optimism and progress. In a society like the GDR, one which prematurely claims for itself the realization of such values, such an aesthetic becomes an apology in its very form. It asserts a false consensus. Its preformulated assumptions disallow *a priori* any expression which would challenge the collective peace. Müller's search for a new aesthetic is also a search for a new politics: instead of sovereignty he calls for doubt; in place of

[19]See David Bathrick, "Reading Walter Benjamin from West to East," *Colloquia Germanica* 12 (1979), 246–255.

[20]Heiner Müller, *Theaterarbeit* (West Berlin: Rotbuch, 1975), p. 125.

optimism, the expression of fear; in lieu of a collective ground in false consensus, the uncompromising voice of the individual.

Not surprisingly, Müller's differences with the Brechtian avant-garde have reached their sharpest contours around the subject of fascism. Where Brecht's enlightened approach would belittle the impact of myth and irrationalism and offer as an antidote the voice of reason, Müller's self (voice) is the fantasy that reason would sublimate and repress.[21] His increasing fixation in his history plays with blood, brutalizing sexuality, and even cannibalism seeks to unveil the silences of an *unbewältigte Geschichte* (a past not yet overcome) whether they reside in the mythologies of Greek, Renaissance, classical, or Marxist-Leninist representations. To this extent Müller also functions as an historian, albeit of a social amnesia. In his play *Germania* (1971) a character called the skull salesman, digging in a cemetery, says it directly:

> I am one of those left over. I am replanting here . . . under flowers and grass. We work nights . . . because of infection. I was an historian. A mistake in periodizing, the Thousand Year Reich, you understand. Since history directed me to the cemeteries, to its theological dimensions, so to speak, I am immune to the cadaver smell of promises for a better world. The golden age lies behind us.[22]

Müller's negative dialectics of history bring to the surface a cultural heritage of the *Deutsche Misere* which has been repressed in the annals of official Marxist historiography. Instead of resistance fighters and positive heroes, he gives us soldiers who cannibalize each other for survival; in place of the birth of the new society on the trail of the victorious Red Army, the grotesque painful rebirth of the old. An example from his play *Germania:* impregnated by Hitler, the madonna Joseph Goebbels gives birth to a horrible thalidomide Wolf.

The artistic tradition that Müller draws on here is clearly of Western and not Eastern origin, of Artaud, existentialism, Samuel Beckett; in short, "cosmopolitanism." Thus it is not surprising that Eastern and Western Marxists alike have accused him of a philosophical pessimism and an individualism associated with such thinking. What their critique fails to consider is the function of such an aesthetic within the "horizon of expectancy" in the GDR. Müller's cross-cut reading of German history pushes against the grain of official representations into distorted fantasies, perverted memories, sordid images—the untold stories. What this in turn entails is the breaking with an aesthetic narrator who posits reality

[21] For an excellent comparison of Müller and Brecht see Helen Fehervary, "Enlightenment and Entanglement: History and Aesthetics in Bertolt Brecht and Heiner Müller," *New German Critique* 8 (Spring 1976), 80–109.

[22] Heiner Müller, *Germania Tod in Berlin* (West Berlin: Rotbuch, 1977), p. 57.

in a subliminated enlightened epic voice. Again Brecht is the target when Müller says, "It is no longer permissible *not* to talk about oneself when one writes."[23] Müller's "self" in plays such as *The Battle* or *The Hamlet Machine* is a repressed male history. His images indeed are not those of resistance fighters or emancipated women or even, as in Brecht, of schizophrenic characters as representations of historical contradictions (Shen Te and Shui Ta of *The Good Person of Sezvan* postulating use value and exchange value; Galileo caught in the imprisonment of false and real consciousness). In Müller—dredged deep from the history of sublimated contradictions—we find the hidden psychic wreckage which enlightened, documented, proletarian history has left behind.

In his book *Male Fantasies* (*Männerphantasien*),[24] Klaus Theweleit has argued for an expanded analysis of fascism beyond the sociopolitical which would grasp it in relation to "a particular male/female relationship (the patriarchal) as the producer of a life-destroying reality." The power of Müller's narration lies precisely in its production of fantasy below and beyond the level of reason. It is below the level of reason for those intellectuals who, like Brecht, have squashed that history into the narrow blocks of a rational consciousness; one that, in the words of Hegel, begins its work after the shades of night have fallen—outside and repressive of the gory underlife. But it is also beyond a level of reason in that Müller is able to catch the living fantasies of those who did fight and die and fantasize as a part of the war machine. We may call this technique anti-patriarchal because it dares to constitute as the imaginary self of dramatic representation the repressed and perverted fantasy life of one group of historical males—a fantasy life which Müller believes to be still alive and well on both sides of the Elbe.

Here Müller's subjectivity and notion of narration come closest to novelist Christa Wolf. With all their differences in temperament and style, both seek the real and a transformation of the real at the level of imaginary production. Although from different genres and experience, both come to terms with patriarchy in part by overcoming the authority of Brecht.

And what does this say about the avant-garde in the GDR today? In conclusion, the following:

(1) That the politicizing of literature within the *Kulturpolitik* guarantees the continuing role of the avant-garde as a political one, even when the target of revolt is one wing of the historical avant-garde itself.

(2) That the problem of subjectivity within socialism must also be ultimately contended with at the level of epistemology and linguistic

[23] *Die Zeit*, March 17, 1978.

[24] Klaus Theweleit, *Männerphantasien* (Frankfurt am Main: Roter Stern, 1977).

representation and for that reason the avant-garde has an important and viable function in exploring buried histories and hidden myths.

(3) That the look to "irrationalist" thinkers such as Lautréamont and Artaud by the left avant-garde in the GDR is connected to the failure of rationalist and objectivist approaches within Marxism to grasp basic truths about human behavior and motivation. "The trends and drives in the historical process," says Müller, "which are above all important for mass movements, have been abandoned to the right wing. I believe we must move away from this kind of overrationalizing. This is why Artaud is a productive nuisance."[25]

(4) Finally, that judgments about the political integrity of any literature can be made only in relation to its function within a particular social order. What serves as individualism or irrationalism in the West takes on a very different character in the GDR. Again Heiner Müller: "I/GDR cannot write about me without writing about GDR/ politics."[26] The realm of intimacy has perforce been politicized. In the case of the poet, he says, this is a burden and a hope.

[25]Interview with Heiner Müller in *Basis: Jahrbuch für Gegenwartsliteratur 6,* edited by Reinhold Grimm and Jost Hermand (Frankfurt am Main: Suhrkamp, 1976), p. 61.

[26]*Die Zeit,* March 17, 1978.

Explosions in the Swamp: Junger's Worker (1932)

JOST HERMAND

> For I see with different eyes than Marx, who discerns a play of economic forces [in everything], while I believe that something more than economics is behind it all.
>
> *Ernst Jünger, 1930*

If we examine the early writings of Ernst Jünger with an eye towards their ideological assumptions, it never ceases to be amazing how simplistic everything is. In spite of all his claims to offer his readers only personal, absolutely "authentic" experiences, the overtones on almost every page are of Nietzsche, Nietzsche, Nietzsche. As is so often the case in literature written by bourgeois intellectuals, experiences which are purely a matter of education and acculturation are passed off here as chthonian, "primeval" experiences. Above all, this man seems to be taken with *Zarathustra,* which so many German soldiers headed for the front dragged around in their packs between 1914 and 1918. Thus, even though Jünger's *Storms of Steel* (*Stahlgewitter,* 1920) often appears to be quite objective and even disinterested, its chief metaphor relies on the concept of "life as a struggle." The real experience of the front becomes transformed into an experience of competition, of *agon,* of war in its pure essence. A battlefield of passions is created here by the confrontation with elemental forces, and it is the "will to power" which decides the outcome. Here, the "greatness" of man (who "is born to be a warrior," according to Nietzsche) is demonstrated by ecstasy, fanaticism, the search for danger, courage, the willingness to sacrifice, suffering loneliness, and, finally, by "tragedy." Indeed, even Nietzsche's aesthetic view of this agonizing struggle[1] is not foreign to Jünger, the platoon leader and wearer of the *pour-le-mérite.* For him, the battlefield is constantly transfigured into an image of painfully convulsing life: a Dionysian experience of chaos, ecstasy, and "authenticity," where, in the midst of apparent meaninglessness, "life *an sich*" is preserved as the supreme and ultimate value which must be maintained, strengthened, and steeled at all costs.

This was the feeling of many Nietzsche-supporters; *Gymnasiasten,* who had become wild and unruly; fanatical disciples of volkish ideology; and other reactionaries who view themselves as outsiders in the "profiteers' republic" of Weimar—the world of bourgeois liberalism and Social Democratic reformism. What is so remarkable or so original about this? Why was it precisely Jünger who became such an important figure for the "right"—not merely as an author, but also as an ideological leader? For after all, Nietzsche was a model for many people at that time who saw the Weimar Republic as only a shabby, transitory phase on the road to a "Third Reich" of renewed German greatness.

Above all there were two things which distinguished Jünger from other reactionaries after the middle of the twenties and which made him into a sensational "case": 1) his thoroughly "revolutionary" nationalism, which made it possible for him to appear as an exponent of the volkish *Freikorps* and 2) his concept of a New Technology and a New Worker, which was influenced by the "New Objectivity." At first sight, both of these ideas seem to be absolutely opposed to Nietzsche's world of thought.

First, there is Jünger's unbridled chauvinism, which promotes "being German" as the supreme value and, indeed, simply equates "being German" with "being authentic" and "being vital." It was this chauvinism which led him to become an active contributor to nationalist journals such as *The Banner* (*Die Standarte*), *Arminius, The New Generation (Die Kommenden),* and *Resistance (Widerstand).* But Nietzsche always shines through, even in this "Soldier's Nationalism"[2] that seems to have emerged from the spirit of the First World War. German warfare is understood here as a strengthening of the one "Great Life" and not tied to

[1]Cf. Friedrich Nietzsche, *Werke,* II (Leipzig: 1910 ff.), p. 52.

[2]Cf. above all Karl Prümm, *Die Literatur des Soldatischen Nationalismus der zwanziger Jahre* (Kronberg: Scriptor, 1974); Prümm, "Vom Nationalismus zum Abendländer. Zur politischen Entwicklung Ernst Jüngers," *Basis,* 6 (1976), 7–29; and Helmut Mörchen, *Schriftsteller in der Massengesellschaft* (Stuttgart: Metzler, 1973), p. 76ff.

a particular party or organization. Thus, Jünger had clear reservations about the *Stahlhelm,* as he had later about the NSDAP. While these organizations steered a "legal" course in the republic, he tended to sympathize with putschists, bomb-throwers, and aristocratic secret societies, whose goal was not so much the welfare of the nation as an affirmation of "wild life." For Jünger, there was nothing to be reformed in the Weimar Republic. For him, it has an "atmosphere like a swamp, which can be purified only by explosions."[3]

The same is true for his concept of the "worker," which at first seems to be as un-Nietzschean as possible. How can such a lowly figure be set up as an ideal typological model on the foundation of such an aristocratic world view? But this is exactly what Jünger does in his book *The Worker: Power and Gestalt* (*Der Arbeiter. Herrschaft und Gestalt*), which appeared in 1932 and was immediately reprinted three times. He was fascinated above all by this figure's harshness, coarseness, and rebelliousness—that is, by its "anti-bourgeois" nature. Therefore, Jünger's "worker" is not determined by rank, class, or sociological factors; but rather—following the typology of *Geistesgeschichte* from the early twenties—it is a *Gestalt,* a "type."[4] In contrast to this literary ideal type, the "bourgeois"—as the counter-image—is accused of all imaginable vices. For Jünger, the "bourgeois" are those who are satiated, comfortable, soft, bored, evasive, and uncombative. They are the philistines who risk nothing, who carefully shield themselves from every danger, and who constantly delve into their own souls. They are the representatives of aimless pluralism, spineless pacifism, and a typically liberal lack of commitment. In short, they are Nietzsche's "last people" who no longer can muster up a sense for "fate," "struggle," "danger," and the "elemental" (p. 46).

In contrast, Jünger views the "worker" as a type with a "sense for struggle" who scorns "pleasures" and possesses an infallible "feeling for manly unconditional systems of value" (p. 235). Upon close scrutiny, his "worker" is not a worker at all, but a descendant of those mercenaries and condottieri who so excited Jünger in the early twenties.[5] His "worker" is the typical Nietzschean warrior who has gone through the experiences of the "New Objectivity" and Italian Fascism, and who has been transformed from a soldier on the battlefield of war to a soldier on the battlefield of production. He is the man of the new harshness who contempt-

[3] *Der Arbeiter. Herrschaft und Gestalt* (Munich: Beck, 1932), p. 245. Quoted subsequently in the text.

[4] Cf. Jost Hermand, *Synthetisches Interpretieren* (Munich: Nymphenburger, 1968), p. 47ff., on these typologies and the polarities of *Geistesgeschichte.*

[5] Prümm, *Soldatischen Nationalismus,* II, p. 401ff. Cf. also such statements by Nietzsche as "Let your work be a struggle . . ." (*Werke,* VI, p. 67f.).

uously rejects the world of humanism, bourgeois softness, and dishonorable peace (the "humiliation of Versailles"), and views struggle and work as the supreme values of a life worth living. For Jünger, this "worker" who lived so dangerously is not an individual, but a type—that is, the *Gestalt* of a new totality; a human being beyond party differences who is distinguished by a "close relationship to the elemental powers" (p. 17), who thinks "heroically" (p. 162), who possesses the "will to power" (p. 67), and who is "committed to the continuation of war" for which Jünger longs (p. 157).

Seen in this way, his "cult of the Germanic" has little to do with the real Germany of 1932, and his concept of the "worker" little to do with the real worker of 1932. Strictly speaking, both are only concepts which are meant to contribute to the strengthening of the life force in the Nietzschean sense. And the "utopia" which Jünger sketches in his book on the *Worker* also remains extremely abstract.[6] In this high-flown manifesto, what he sees through the gunsights as the world of tomorrow is an infinite landscape of technology, a gigantic planned economy, and a non-bourgeois life consisting of prodigious efforts to develop armaments—since Jünger, like Nietzsche, is convinced that war is the most elemental force of nature and that the human race renews itself only through constant competition. Therefore, he places his only hope for the future on an age of great wars, and not on the bourgeois dream of "Eternal Peace." And he finds these hopes thoroughly vindicated by economic development, intensified competition, imperialism, and the severe economic crisis after 1929.

Jünger sees that the wars between modern, highly industrialized states are becoming more and more murderous. However, this does not fill him with anxiety, but with confidence. He knows no "bourgeois" horror of ecological crises, overpopulation, devastation, and advancing technology. Quite the contrary: after a "bourgeois" phase of satiation, he sees clear signs in these developments of an intensification of the crisis and thus of a strengthening of the will to life in those industrious, strong, combative natures who, in the face of the confrontations to come, re-embrace harshness, cruelty, irrationality, and even life-giving barbarism. And Jünger also sees that it will no longer be muscular strength or individual prowess which will be decisive in this future "age of wars" that Nietzsche had already dreamed of; rather, it will be exclusively a matter of material superiority, of "firepower." Hence his cult of technology, of planned labor, of the "worker." Jünger's worker, who surpasses even the warrior on this imperialistic battlefield of production, is always the maker

[6]On Jünger's anti-humanistic "utopia" of nihilism, cf. Gerhard Loose, *Ernst Jünger. Gestalt und Werk* (Frankfurt: Klostermann, 1957), p. 120.

of armaments, the soldier-worker, the fighter at the place of production, the type of the heroically steeled engineer, great and relentless, the industrial Superman—and no real proletarian.

Even Jünger's cult of technology, which would have been totally foreign to Nietzsche, serves the same Nietzschean purposes: not the bourgeois "victory of technology" but the creation of a post-bourgeois "chaos," of an irrational world of fiery abysses and explosions in which there is no real victory but only destruction tinged with a heroic glow. In this respect, rather than allying himself with such anti-war, pacifist authors as Arnold Zweig, Erich Maria Remarque, Theodor Plivier, Ludwig Renn, or Edlef Köppen, Jünger prefers to side with Oswald Spengler, who in his book *Man and Technology: Contributions to a Philosophy Life* (*Der Mensch und die Technik. Beitrag zu einer Philosophie des Lebens,* 1931), had represented Nordic man as the most ingenious beast of prey, and world history as a succession of more and more violent catastrophes. After all, Spengler had written à la Nietzsche: "Peace, happiness, and pleasure are unknown to the highest forms of the species."[7] And, he had termed fatalism the only *Weltanschauung* "which is worthy of us." "Rather a short life full of deeds and fame," he writes, "than a long life without meaning."[8]

And so, is there any sort of future or any sort of state in Jünger's "utopia?" And how are social relations to be organized within it? In this respect, too, Jünger proves to be absolutely unoriginal. On the one hand, like Nietzsche, he totally rejects the state and emphasizes only the strengthening of the race, of blood, of male resolve, of "life." On the other hand, like the George Circle or the *Jungdeutscher Orden,* he dreams of a system of "mastery and service" in which those who command as well as those who obey place themselves completely in the service of the whole. Here, all people—the leaders and the workers—are the "foremost servants of their state" (p. 13). That is, Jünger is thinking of a state constructed along the lines of the "Teutonic Order, the Prussian Army, or the Society of Jesus" (p. 201), in which the ruling principle is "all for all." Instead of resting content with a peaceful "world of ants," as the bourgeois liberals do, he dreams of "imperial, planned landscapes" (p. 292) which are no longer dominated by the happiness of the broad masses, the common herd, and the rabble, but by the male, warlike concept of "harshness." Here, the worker-warrior dedicates himself "to steeling his weapons and his heart in the midst of chaotic zones" and learns to renounce the "temptation of happiness" (p. 292). Nietzsche had

[7]Oswald Spengler, *Der Mensch und die Technik. Beitrag zu einer Philosophie des Lebens* (Munich: Beck, 1931), p. 57.

[8]*Ibid,* p. 88.

already written: "What do I care for happiness? I care only for my works."[9] Therefore, Jünger's social goal is a kind of workers' aristocracy which, in full possession of power, is totally devoted to its heroic tasks of arms mobilization and waging war. In contrast, as was already true for Nietzsche, those who really work are belittled as the masses, as coolies, as an inferior "type of Chinaman."[10]

Most of these ideas are either vulgarized Nietzsche or the heroically exalted enthusiasm for the *Bund,* which had many supporters among bourgeois youth after 1929. Moreover, this senseless utopianism is influenced by those concepts of Fordism or Taylorism which were part of the "New Objectivity"[11] and the "White Socialism" proclaimed by Spengler as early as 1921 in his book *Prussia and Socialism* (*Preussentum und Sozialismus*). Spengler had already described the French as "citizens" and the Germans as "workers."[12] Spengler had already characterized German socialism as an "authoritarian, illiberal, and anti-democratic" socialism whose power emanates from the whole. Spengler had already labeled the officer and the worker as the best Germans, because they were the most Prussian. Spengler had already invoked "fate," "blood," and the tradition of the "Teutonic Order."[13] Spengler had already seen the Germanic "will to power" in Prussian socialism—that is, in non-Marxist socialism. Spengler had already addressed the "elect among German workingmen" and the defenders of "traditional Prussian national sentiments" in the following words: "We need harshness; we need courageous skepticism; we need a class of socialists who are masters by nature. Again: socialism means power, power, and more and more power."[14]

Jünger propagates the same naked, rapacious imperialism. Therefore, his socialism does not see its real enemy in capitalism, but in "false"—that is Marxist—socialism. Upon close inspection, Jünger's "socialism" proves to be only an imperialistic capitalism: a heroic order of warriors in technological disguise, committed to the principle of "total mobilization," "the total state," and "total war" with regard to the final struggle for control of the world. Jünger's future state is a symphony in steel which renounces all the medieval romanticism of Stefan George's followers or the blood-and-soil fanaticism of the Fascists and shamelessly embraces the most modern machinery and, above all, the most advanced weapons. In contrast to all ideals of pre-industrial backwardness, this

[9]Nietzsche, *Werke,* VI, p. 476.
[10]*Ibid,* VIII, p. 153.
[11]Helmut Lethen, *Neue Sachlichkeit 1924–1932* (Stuttgart: Metzler, 1970), p. 91f.
[12]Oswald Spengler, *Preussentum und Sozialismus* (Munich: Beck, 1921), p. 10.
[13]*Ibid,* pp. 42, 53.
[14]*Ibid,* p. 98f.

emphasis on avant-garde technology gave his "utopia" an extremely seductive appeal which did not fail to have an effect. For here, all sentimentalities had been renounced. Here, even Nietzsche's ideal of love for the man to come, the Superman who was to be created, had been abandoned. Here, everything was merely cold steel, the most blatant struggle for existence, chilling danger, and heroic nihilism.

These ideas of the *Total Mobilization* (*Totale Mobilmachung,* 1930) and the *Worker* (1932) found their most enthusiastic reception among those "Leftists of the Right" who were committed to the "Conservative Revolution" and who increasingly opposed the Weimar Republic after 1929 by advocating concepts of youth, community, and the Germanic character.[15] All of these circles viewed the bourgeois-capitalist system as something which had been totally ruined by mismanagement and which could not be reformed at all. Indeed, not only capitalism, but all traditional ideologies—whether conservatism, liberalism, Social Democracy, or socialism—seemed to express the same *misère:* namely, a pluralistic understanding of the world which granted a political voice even to the most ignorant and the weakest members of society. These circles saw communism as thoroughly "compatible with the system," since they believed it encouraged the working class to adopt bourgeois values and thus to sink in the same swamp. In order to overcome the universal crisis, they demanded a "fundamental revolution": that is, a path between the KPD and the NSDAP, a path to a National Bolshevism founded on the principle of the *Bund.* At the time, this was advocated in the most effective manner by the circle around Hans Zehrer in the journal *The Deed* (*Die Tat*) and the circle around Ernst Niekisch in the journal *Resistance* (*Widerstand*).

Niekisch was probably Jünger's most ardent admirer, having just published his own book *Adolf Hitler: A German Fate* (*Adolf Hitler. Ein deutsches Verhängnis*) in 1932 and having praised Jünger's *Worker* in *Resistance* as the greatest, most important, and most significant book of the year. According to Niekisch, Jünger's book advocates "a philosophy of German Bolshevism which is free of all Marxism"; it highlights the "imperial rank" of the worker; and thus it deals a death-blow to the "bourgeois-capitalist world."[16] By helping Germany in the struggle "against the West, against Versailles," we read here, it supports the same "general tendency" as "Bolshevist Russia."[17]

[15] Cf. Otto-Ernst Schüddekopf, *Linke Leute von rechts* (Stuttgart: Kohlhammer, 1960), p. 258ff.

[16] Ernst Niekisch in *Widerstand,* No. 8 (1932), 257.

[17] *Widerstand,* No. 10 (1932), 307, 310f. Niekisch maintained this perspective even after 1945 and continued to praise the *Worker* for its "socialism of the front" and its "National Bolshevist" tendencies. Cf. *Antaios,* 5/6 (1965).

But such a "National Bolshevist" interpretation was probably only wishful thinking. Jünger's actual intention in this book aims far beyond National Bolshevism, National Socialism, and the "Third Way" of the circle around *The Deed*. In contrast to these groups, he did not want to resolve the growing crisis. Quite the contrary: he wanted to intensify it in order to have the contradictions become even more more blatant. He wanted to increase the "chaos." His goal, like Nietzsche's, was substantially greater, more abstract, more aesthetic, and more inhuman: it was the catastrophe, the tabula rasa, and heroic, total destruction. Jünger did not want to offer a helping hand. "Whatever falls should also be kicked"—that was his maxim, too.[18] Hence his "pathos of distance" towards all parties, even the NSDAP.[19]

Such a viewpoint does not relegate Jünger's *Worker* to the aesthetic or literary realm. In such matters it is not so much the subjective intention which is decisive as the objective goal and the objective effect. And this effect—like that of Spengler and Nietzsche between 1929 and 1932—was to provide gratuitous propaganda for patriarchy, the cult of the *Führer*, and obedience to authority, and thus to support racism, capitalism, and imperialism.[20] Indeed, in the long run it also served to make the NSDAP palatable to certain intellectuals and to legitimize this party as representing a "New Way." But not only that. When all is said and done, Jünger's ideology was more merciless, more bestial, and more Nietzschean than that of the Nazis, for he was openly steering toward chaos and the end of the world—with the help of the most modern systems of weapons. Compared to this, even fascism seems almost harmless—at least in its early years. No wonder Jünger maintained an aristocratic aloofness in 1933 from such a "petty bourgeois" doctrine of salvation. He had hoped for worse things.

After the collapse of the Third Reich, Jünger was not entirely sure whether he should reissue his *Worker*. It was only the persistent urging of Martin Heidegger which persuaded him to include it in its original version in the sixth volume of his collected *Works*. And the club of the initiated thanked him for this, especially in the "evil" sixties, when the rejuvenated Left began a consistent attempt to "overcome the past" and Jünger's star gradually began to grow dim. Unfortunately, these times are over in the Federal Republic. When Jünger received the prestigious Schiller Prize in 1975 on the occasion of his eightieth birthday (after a short, dangerous life full of peril and sacrifice), he was congratulated by

[18] Nietzsche, *Werke*, VI, p. 305.

[19] *Ibid*, VIII, p. 436. Cf. also Jünger's essay "Ueber den Schmerz" (1934), where he not only uses Nietzsche's expression: "pathos of distance," but also speaks of the "will to power" and the "last people." In *Werke*, V (Stuttgart: Klett, 1960), pp. 161, 181.

[20] Cf. also Georg Lukács, *Die Zerstörung der Vernunft* (Berlin/DDR: Aufbau, 1955), pp. 422ff.

an immense chorus of well-wishers. Today, even his *Worker* is extolled in all candor as an "exciting, furious, and colossal book"; and it is placed on the same level with such works as the *Communist Manifesto* and Nietzsche's later writings, as we can read in a long essay by Gerd-Klaus Kaltenbrunner in the supposedly "liberal" *Frankfurter Allgemeine Zeitung* of 2 September 1977.[21]

Such people know what they are doing. In spite of all its confusion and perverseness, even a book like this is still good enough to be used as a defense of the capitalist system, which must constantly legitimize itself in the face of opposition from the Left. Kaltenbrunner is playing with a relatively open hand. Above all, he praises Jünger's *Worker* for its new "order" (read: the "well-regulated society"), its "aggressive fatalism" (read: the renunciation of perspectives for social change), its elevation of the worker into the "metaphysical realm" (read: the return to religion), and its renunciation of "happiness" (read: the old slogan of thriftiness). We read here that Jünger's "Superstate" is based on "authority, strong bonds, discipline, and breeding"; that is, on the "primacy of elite, heroic, and imperial virtues." And Kaltenbrunner finds this to be thoroughly acceptable. Therefore, he in no way sees Jünger's vision of a "planetary dictatorship" of technocrats (read: rule of the "multi-nationals") as "antiquated, but rather as closer to reality than it was forty-two years ago." This could not be said in a more blatant way. Such statements can be outdone only by Jünger himself, who is said to have once claimed: "With respect to historical reality, I am ahead of the times—that is, I perceive things somewhat earlier, somewhat before they appear."[22] Let us hope that someday this "prophesying" will be exposed as the arrogant ramblings of a near-sighted, minor visionary!

Translated by Carol Poore

[21]Gerd-Klaus Kaltenbrunner, "Konservative Apokalypse. Wiedergelesen: Ernst Jüngers *Arbeiter* (1932)," *Frankfurter Allgemeine Zeitung*, 2 Sept. 1977, 23.

[22]Quoted in Kaltenbrunner.

III
Science Fictions

Introduction

TERESA DE LAURETIS

They don't make futures like they used to.

Within the last decade academic curricula and literary criticism have become very much concerned with science fiction; at the same time there has been a growing interest in science fiction on the part of writers and readers. These interconnected events—the increased production, intensified reception, and wider circulation of science fictional texts—have generated a series of discourses *on* science fiction which takes its place alongside the discourse *of* science fiction and the many other practices and representations that shape the contemporary social imagination.

The purpose of this section is to explore some of the theoretical conditions and historical aspects of a critical discourse on science fiction (SF).

The "traditional" concerns of SF, as they were established in American pulp magazines of the 1920s and 1930s, remain very much alive—the concern with scientific or cognitive processes, with problem solving and ludic extrapolation, with futurological if not predictive speculation in the direction of alternative reality structures and social visions. But a shift in emphasis has taken place.

From the physical and biological or hard sciences and the corresponding technological hardware, SF has turned to what may be called the sciences of the symbolic—linguistics, communication theory, economics, semiotics, symbolic logic—i.e., those methodologies that have been developing in direct relation to historical and social phenomena. More significantly, as the culture's attitude toward science has changed, so have the epistemological premises of SF: no longer accepted as an *explanation* of natural or social phenomena, whose validity is directly proportional to its predictive or normative capacity on a *universal* level, science is understood as, itself, a set of *historical* and culture-bound discourses on material reality, physical and/or social.

Whether this is due to a more correct understanding of the scientific enterprise on the part of a better educated public; or to the process of de-ideologization of science and technology that was initiated in the 1960s in a climate of intensified political awareness; or to other factors involved in the shift, this is certainly a fundamental concern of SF criticism. And even though such considerations exceed the limitations of this volume, it is important to delineate, however broadly, some aspects of the historical context of contemporary SF in the United States. Thus one must point, minimally, to a series of social practices that have been developing since the 1960s, when the public began to question the actual impact of the scientific-military-industrial complex on social life—practices that have led to the current public interest in ecology, human rights, racial and sexual difference, alternative technologies, physical and psychological survival strategies, and the development of human resources.

This more direct relation of the people to the environment and the concomitant belief in collectively determined, community-oriented, social action are envisaged in the possible worlds of contemporary SF. It is no longer haunted by the dystopian specter of an all-powerful, uncontrollable technology which takes over the whole world and reduces humans to alienated, undifferentiated cogs in its totalitarian mechanism (echoes of the purges of the 1950s, the specter of Communism, the cold war, etc.). It more often presents technology as a form of techne, creative making, practical knowledges, immediately useful activity for people's daily existence, basic needs *and pleasures*—physical, aesthetic, and intellectual. Even when the possible world constructed is a galactic confederation or a bureaucratic empire, technology is part of the human landscape, affecting both individual and cultural practices such as cosmetic surgery or the nurturing and childbearing capacity, artistic performance or food growing. Thus it entertains a double and equally important relationship to fantasy and subject processes, and to labor and social reproduction. The most striking aspect of such technology is neither its power to produce (its value as a perfect machine, as Capital) nor its products (the exchange value of the commodities it produces) but precisely its use value for those who develop it and use it, who are engaged in the process of making, creating, producing socially useful objects, knowledges, and pleasure.

The assimilation of technology and art as socially useful forms of poietic activity in this science fictional vision reflects and is reflected by social practices like super 8 home movies, mimeographed fan publications (fanzines), CB radios, or community street festivals. Just as it is useless if not impossible to judge such events by high art or museum standards, so it is to judge SF writing by the standards developed in other historical periods for those periods' literary productions. Other standards of artistic value, other critical methodologies, other aesthetic canons must be developed in relation to all those expressive-communicative phenomena which, like SF, can neither be categorized as literature or art in the traditional sense nor simply lumped under the pejorative label of popular culture. On the contrary, each such phenomenon must be viewed as having a generic specificity and a history, inscribed both in the concrete social practices and in the universe of theoretical discourses which constitute the conceptual horizon, the speculative vision, the social imagination of the culture in which it is produced.

Let's take, for example, SF's extraordinary concern with language. It is obviously related to the craft of writing and the vast body of written literature on which SF necessarily draws. It is also related to the actual social practice of dialects, slang, technical or specialized usage, particular syntactic or narrative patterns, and more or less conventionalized paralinguistic behavior. But a third, theoretical link must be established: language as the primary, universal example of human symbolic behavior is also one of the focal points of Western thought since the beginning of this century, as witness the development of modern linguistics, anthropology, economics, psychoanalysis, semiotics—the very sciences which constitute the present horizon of SF.

Language, however, is the stuff of all literary work, and at all times writers have been supremely concerned with it as all artists are with their medium and the matter of expression. This, then, makes it necessary, in speaking of SF as a form of verbal art, to examine the following interrelated questions:

(1) the historical specificity of SF as an aesthetic production, its modes of address, its reception, and its relation to the public sphere; and

(2) the generic question, SF's particular deployment of language and cognitive processes in relation to other forms of fiction and literary writing.

The three essays in this section address both of these questions, from different but not inconsistent angles or points of entry. Within a historical materialist critical perspective, their theoretical assumptions range from Hegel and Marx to Bloch, Eco and Foucault. The three authors share a view of SF as a specific organization of textual elements and strategies mediating concrete historical and social relations; but they differ in their understanding of textuality, of the ways in which the text produces and is produced by those mediations. Two views of the text operate in this section, which thus also poses the problematic of textuality and the terms of a debate, not only literary but epistemological, about the very conditions of representation in SF and in the critical discourse on SF.

One view posits the text (the book, the story, the novel) as a coherent whole held together by an internal unifying agent or device not unlike the Aristotelian unities, and containing well-defined meaning(s) that are shared, and exchanged, by author and reader. The critic's goal is to expose the device, to integrate the elements and levels of the work, finally to reconstitute it as a totality of meaning. This is a structural (if not structuralist) approach. The other, which I would call semiotic, sees the text not as a product of cognition but as the locus of its processes, not as the blueprint of an ultimate signified but as a mapping of uncharted signifiers into labile and shifting, always partial meanings; in this view, the text is precisely what allows meaning to be displaced and deferred, not localized or permanently attained in either writing or reading. The critic's activity is then to trace possible paths of reading, to open gaps in the presumed coherence of discourse and make room for the play of language and the plurality of meaning, to propose implicative rather than explicative models of mediation.

Darko Suvin's "Science Fiction and the Novum" offers a theoretical definition of the genre based on what he sees as the *differentia specifica* of SF: "the narrative dominance or hegemony of a fictional novum (novelty, innovation) validated by cognitive logic." Distinguishing SF from other genres—e.g., naturalistic fiction, fantasy, and allegory—Suvin identifies the strategies of the SF narration as an oscillation, generating an irreversible change, between the "zero world" of the author's/reader's known reality and an alternative historical reality; this conceptual movement, predicated on and validated by a post-Einsteinian and post-Marxian scientific method, is made possible by a device or means of reality displacement, the fictional novelty, conceived not as a formal innovation so much as a historical category: "the aesthetic novum is either a translation of historical cognition and ethics *into* form, or . . . a creation of historical cognition and ethics *as* form." Suvin's model is meant to be explicative and normative, especially with regard to the novum's relevance to a positive human development, which he sees as based on "law-like tendencies in men's social and cosmic history."

In "Signs of W$^{a}_{o}$nder" I view the SF text "as a discontinuous set of sign-functions which produce an indefinite set of semantic constructs by dislocating the subject in the reading process"; my model is implicative and intended to disrupt the effects of narrativization by resisting its tendency to totalize meaning. The analytical tool proposed is the notion of *sign-work* which, combining the semiotic notion of sign with the dynamic subject processes of Freud's dreamwork (*Traumarbeit*), is formulated to engage rhetorical and other textual processes that exceed a narrowly conceived linguistic order. The essay argues that the historical and generic specificity of SF, as distinguished from mainstream, modernist or postmodernist fiction, is to be sought in SF's re-literalization of language, its way of posing the relation between human figure and physical/material/technological ground, its ability to reshape the semantic universe, and to produce other conceptual models of human reality. Hence the radical hypothesis that SF today may have assumed the social and aesthetic functions once held by Greek epos and, in the more recent past, by realism.

Samuel R. Delany's "Generic Protocols: Science Fiction and Mundane" reformulates the genre question by describing SF as a reading protocol, a structuration of response potential. Against what he calls "the metafictive fallacy," namely that all writing is an elaboration or a critique of other texts, of former writing protocols, while reading is merely a transparent gaze which connects the written text and understanding (meaning), Delany argues that meaning formation is a function of reading and thus of generic protocols. Reading is "a mental field of response that a history of exposure to various written statements contours toward a structuration which *is* understanding." A genre is a reading protocol, shaped historically by a set of texts and on which new texts intrude continually changing it. And writing, therefore, is a critique of current reading protocols, each text recontouring the response potential and and reshaping the genre. Because of the growing interest in SF, a rupture and an encounter have taken place between SF and *mundane* fiction (as fans call literature and all non-SF stories set in this/the world, in the present or past); and the battlefield is, precisely, that of literary criticism. Here, Delany shows, the evaluative standards traditionally dependent on theoretical, stylistic or historical unity and centered on the author-function must yield to the fundamental plurality of the SF protocols and to values of disunity, heterogeneity and play that expand, rather than totalize, the interpretive space created by the text.

SF and the Novum*

DARKO SUVIN

To the memory of Ernst Bloch

0. It is often thought that the concept of a literary genre (here science fiction, further SF) can be found directly in the objects investigated, that the scholar in such a genre has no need to turn to literary theory since he/she will find the concepts in the texts themselves.[1] True, the concept of SF is in a way inherent in the literary works—the scholar does not invent it out of whole cloth—but its specific nature and the limits of its use can be grasped only by employing theoretical methods. The concept of SF cannot be extracted intuitively or empirically from the works labeled "science fiction." This is still often tried by positivistic critics; unfortunately, the concept arrived at in such cases is primitive, subjective, and unstable. In order to determine it more pertinently and delimit it more precisely, it is necessary to educe and formulate the *differentia specifica* of the SF narration. My axiomatic premise in this essay is that *SF is distinguished by the narrative dominance or hegemony of a fictional novum (novelty, innovation) validated by cognitive logic*. Let me try to go through this premise at some length and attempt to justify its use by the consequences it entails.

*This essay, now also published in Suvin, *Metamorphoses of Science Fiction* (New Haven, Ct. and London: Yale Univ. Press, 1979), was first presented as a paper during the International Symposium on Postindustrial Culture sponsored by the Center for Twentieth Century Studies at the University of Wisconsin-Milwaukee in November 1977. At that time, the author granted permission for its publication in this volume. (The Editors)

[1]This article is the outgrowth of experiences acquired in two research projects, in which I gratefully acknowledge the financial aid of a Québec Ministry of Education FCAC grant and of a Canada Council research grant.

1. *The Novum and Cognition*

1.1 What is the common denominator whose presence is logically necessary and which has to be hegemonic in a narration in order that we may call it an SF narration? In other words, how can the proper domain of SF be determined, on what theoretical axis does it hinge? The answer is clouded by the present wave or irrationalism, engendered by the deep structures of the irrational capitalist way of life which has reduced the dominant forms of rationality itself to something narrow, dogmatic, and sterile inasmuch as they are the forms of reasoning of the dominant or of the dominated classes. Nonetheless, I do not see any tenable internal determination of SF which would not hinge on the category of the *novum*, to borrow (and slightly adapt) a term from the best possible source, Ernst Bloch.[2] A *novum* or cognitive innovation is a totalizing phenomenon or relationship deviating from the author's and implied reader's norm of reality. Now no doubt, each and every poetic metaphor is a novum, while modern prose fiction has made new insights into man its rallying cry. However, though valid SF has deep affinities with poetry and innovative realistic fiction, by "totalizing" I mean a novelty entailing a change of the whole universe of the tale, or at least of crucially important aspects thereof. As a consequence, the essential tension of SF is one between the readers, representing a certain number of types of Man of our times, and the encompassing and at least equipollent Unknown or Other introduced by the novum. This tension in turn estranges the empirical norm of the implied reader (more about this later). The postulated innovation can be of quite different degrees of magnitude, running from the minimum of one discrete new "invention" (gadget, technique, phenomenon, relationship) to the maximum of a setting (spatiotemporal *locus*), agent (main character or characters) and/or relations basically new and unknown in the author's environment. (Tangentially be it said that this environment is always identifiable from the text's historical semantics, always bound to a particular time, place and sociolinguistic norm, so that what would have been utopian or technological SF in a given epoch is not necessarily such in another—except when read as a product of earlier history. In other words, SF is a *historical* genre.)

1.2 The novum is postulated on and validated by the post-Cartesian and post-Baconian scientific *method*. This does not mean that the novelty is primarily a matter of scientific facts or even hypotheses, and insofar as the opponents of the old popularizing Verne-to-Gernsback orthodoxy protest against such a narrow conception of SF, they are quite right. But they are not right when they deny that what differentiates SF from the "super-

[2]In particular, Ernst Bloch, *Das Prinzip Hoffnung* (Frankfurt am Main: Suhrkamp, 1959) and *Experimentum Mundi* (Frankfurt am Main: Suhrkamp, 1975).

natural" literary genres (mythical tales, fairy tales, etc., as well as horror and/or heroic fantasy in the narrow sense) is the presence of scientific cognition as the sign or correlative of a method (way, approach, atmosphere, sensibility) identical to that of a modern philosophy of science.[3] Therefore, science in such a wider sense cannot be disjoined from the SF innovation, in spite of fashionable currents in SF criticism of the last fifteen years, though it should conversely be clear that a proper analysis of SF cannot focus on its explanatory scientific *content*. Indeed, a very useful distinction between "naturalistic" fiction, fantasy, and SF, due to Professor Philmus, is that naturalistic fiction does not require scientific explanation, fantasy does not allow it, and SF both requires and allows it.[4]

In other words, if the novum is the necessary condition of SF (differentiating it from naturalistic fiction),[5] the validation of the novelty by scientifically methodical cognition into which the reader is inexorably led is the *sufficient* condition for SF. Though such cognition cannot, in a work of verbal fiction, be empirically tested either in the laboratory or by observation in nature, it *can* be methodically developed against the background of a body of already existing cognitions, or at the very least as a "mental experiment" following accepted scientific, i.e., cognitive, logic. Of the two, the second alternative—the intrinsic, culturally acquired cognitive logic—seems theoretically the crucial one to me. Though I would be hard put to cite an SF tale whose novelty is not in fact continuous with or at least analogous to existing scientific cognitions, I would be disposed to accept, theoretically, the faint possibility of a fictional novum that would at least seem to be based on quite new, imaginary cognitions, beyond all real possibilities known or dreamt of in

[3]In this essay, "science" is thought of as *sensu lato,* similar to German *Wissenschaft* and French *science,* to methodically systematic cognition. I have tried to explain this further in my essays, "On the Poetics of the Science Fiction Genre," in *Science Fiction,* ed. Mark Rose (Englewood Cliffs, N.J.: Prentice Hall, Inc., 1976), and "'Utopian' and 'Scientific'," *The Minnesota Review,* NS No. 6 (1976).

[4]The distinction is to be found in Robert M. Philmus, "Science Fiction: From its Beginning to 1870," in *The Anatomy of Wonder,* ed. Neil Barron and Peter Doiron (New York: R.R. Bowker Co., 1976), pp. 3–16.

[5]Works avowedly written within a non-realistic mode, principally *allegory* (but also whimsy, satire, and lying tall tale or Münchhauseniade), constitute a category for which the question of whether they possess a novum cannot even be posed, because they do not use the new worlds, agents, or relationships as coherent (albeit provisional) ends, but as immediately transitive and narratively non-autonomous means for direct and sustained reference to the author's empirical world and some system of belief in it. The question whether an allegory is SF, and vice versa, is meaningless. This means that—except for exceptions and grey areas—most of the works of Kafka or Borges cannot be claimed for SF: though I would argue that *In the Penal Colony* and *The Library of Babel* would be among the exceptions.

the author's empirical reality. (My doubts here are not so much theoretical as psychological, for I do not see how anybody could imagine something not even dreamt of by anyone else before; but then I do not believe in individualistic originality.) But beside the "real" possibilities, there exist also the much stricter—though also much wider—limits of "ideal" possibility, meaning any conceptual or thinkable possibility whose premises and/or consequences are not internally contradictory.[6] Only in "hard" or near-future SF does the tale's thesis have to conform to a "real possiblity," to that which is possible in the author's reality and/or according to the scientific paradigm of his culture. On the contrary, the thesis of *any* SF tale has to conform to an "ideal possibility," as defined above. Any tale based on metaphysical wishdreams—e.g., omnipotence—is "ideally impossible" as a coherent narration (can an omniscient being create a stone it will not be able to lift?, etc.), according to the cognitive logic humanity has cumulatively acquired in its culture from the beginnings to the present day. It is intrinsically or by definition impossible for SF to acknowledge any meta-physical agency, in the literal sense of an agency going beyond *physis* (nature). Whenever it does so, it is not SF, but a metaphysical or (to translate the Greek into Latin) a supernatural fantasy-tale.

1.3 Thus, science is the encompassing horizon of SF, its "initiating and dynamizing motivation."[7] As I said, this emphatically does not mean that SF is "scientific fiction" in the literal, crass, or popularizing sense of gadgetry-cum-utopia/dystopia. Indeed, a number of important clarifications ought immediately to be attached: I shall mention three. A first clarification would be that horizon is not identical to ideology. Our view of reality or conceptual horizon is determined by the fact that our existence is based on the application of science, and I do not believe we can imaginatively go beyond such a horizon; today a machineless Arcadia is simply a microcosm with zero-degree industrialization and a lore standing in for zero-degree science. On the other hand, within a scientific paradigm and horizon, ideologies can be either fully supportive of this one and only imaginable state of affairs, or fully opposed to it, or anything in between. Thus, anti-scientific SF is just as much within the scientific horizon (namely a misguided reaction to repressive—capitalist or bureau-

[6] I have been stimulated by the discussion of Ivan Foht, "Slika čovjeka i kosmosa," *Radio Beograd Treći program* (Spring 1974), pp. 523–60.

[7] Jan Trzynadlowski, "Próba poetyki science fiction," in *Z teorii i historii literatury*, ed. K. Budzyk (Warsaw, 1963), p. 272. See also Stanislaw Lem, *Fantastyka in futurologia* (Krakow: 1973); Rafail Nudelman, "Gespräch im Kupe," in *Polaris 2*, ed. Franz Rottensteiner (Frankfurt: 1974), pp. 49–75; and Joanna Russ, "Towards an Aesthetic of Science Fiction," in *Science-Fiction Studies*, ed. R. D. Mullen and Darko Suvin (Boston: Gregg Press, Inc., 1976), pp. 8–15.

cratic—abuse of science) as literary utopia and anti-utopia both are within the perfectibilist horizon. The so-called "speculative fiction" (e.g., Ballard's) clearly began as and has mostly remained an ideological inversion of "hard" SF. Though the credibility of SF does not depend on the particular scientific rationale in any tale, the significance of the entire fictive situation of a tale ultimately depends on the fact that "the reality that it displaces, and thereby interprets"[8] is interpretable only within the scientific or cognitive horizon.

A second clarification would be that *sciences humaines* or historical-cultural sciences such as anthropology-ethnology, sociology, or linguistics (i.e., the mainly non-mathematical sciences) are equally based on scientific methods such as: the necessity and possibility of explicit, coherent, and immanent or non-supernatural explanation of realities; Occam's razor; methodological doubt; hypothesis-construction; falsifiable physical or imaginary (thought) experiments; dialectical causality and statistical probability; and progressively more embracing cognitive paradigms. These "soft sciences" can probably better serve as basis for SF than the "hard" natural sciences; and in fact, they *have* been the basis of all better opuses in SF, partly through the characteristic subterfuge of cybernetics, the science in which hard nature and soft humanities fuse. A third clarification, finally, would be that science since Marx and Einstein is an open-ended corpus of knowledge, so that all imaginable new corpuses which do not contravene the philosophical basis of the scientific method in the author's times (e.g., the simulsequentialist physics in Le Guin's *The Dispossessed*) can play the role of scientific validation in SF.

1.4 It can be objected to this, empirically, that a good proportion of what is sold as SF is constituted by tales of more or less supernatural or occult fantasy. However, this is the result of the ideological and commercial habit of lumping together SF and fantasy (or fiction whose novum *is* and fiction whose novum *is not* cognitively validated) on the purely negative basis of their not being identical to the norm of the author's reality and naturalistic fiction. On a deeper and even more pathological level—internalized in fictional creation—this has led to tales which incongruously mingle science-fictional and fantastic narrative. A misshapen sub-genre born out of such mingling is "science-fantasy," from Robert Chambers (if not Poe) through Merritt to Bradbury; about it I could only repeat after the late and regretted James Blish, that "plausibility is specifically invoked for most of the story, but may be cast aside in patches at the author's whim and according to no visible system or principle," in "a blind and grateful *abandonment* of the life of the mind."[9] In supernatural fantasy proper, the

[8]Robert M. Philmus, *Into the Unknown* (Berkeley: Univ. of California Press, 1970), p. 20.

[9]William Atheling, Jr. (pseud. of James Blish), *More Issues at Hand* (Chicago: Advent Pubs. Inc., 1970), pp. 98 and 104. A further warning in the same chapter that the hybrid of SF and

supposed novelty rejects cognitive logic and claims for itself a higher "occult" logic—whether Christian, a-Christian and indeed atheistic (as is the case of H. P. Lovecraft), or most usually an opportunistic blend of both, openly shown in the more self-confident nineteenth century by Marie Corelli's "Electric Christianity," whose enormous popularity is echoed right down to C. S. Lewis. The consistent supernatural fantasy tale—one which does not use only a single irruption of the supernatural into everyday normality, as in Gogol's *Nose* or Balzac's *Peau de Chagrin,* but develops the phenomenology of the supernatural at the expense of the tension with the everyday norm—is usually (in England from Bulwer-Lytton on) a proto-fascist revulsion against modern civilization, materialist rationalism, and such. It is organized around an ideology unchecked by any cognition, so that its narrative logic is simply overt ideology plus Freudian erotic patterns. If SF exists at all, this is not it.

One of the troubles with distinctions in genre theory is, of course, that literary history is full of limit-cases. Let me briefly examine one of considerable importance, Stevenson's *The Strange Case of Dr. Jekyll and Mr. Hyde.* Despite his literary craftsmanship, Stevenson is cheating in terms of his basic narrative logic. On the one hand, his moral allegory of good and evil takes bodily form with the help of a chemical concoction. On the other hand, not only does the Jekyll-Hyde transmogrification become unrepeatable because the concoction had unknown impurities, but Hyde also begins "returning" without any chemical stimulus, by force of desire or habit. This unclear oscillation between science and fantasy, where science is used for a partial justification or added alibi for those readers no longer disposed to swallow a straightforward fantasy or moral allegory, is the reason for the elaborate, clever but finally not satisfying exercise in detection which naturalistically shelves but does not explain the fuzziness at the narrative nucleus. This marginal SF is therefore, to my mind, an early example of "science fantasy." Its force does not stem from any cognitive logic, but rather from the anguish of Jekyll over his loss of control and from the impact of the underlying moral allegory—the latter being very relevant to Victorian bourgeois repressions of the non-utilitarian or non-official aspects of life, and simultaneously, an unsubstantiated promise that the oscillation between SF and fantasy does not matter since we are dealing with an allegory anyway (see my note 5).

---

detective tale leads—as I would interpret it, because of the incompatibility between the detective tale's contract of informative closure with the reader and the manifold surprises inherent in the SF novum system—to a trivial lower common denominator of the resulting tale, has so far been developed by Rafail Nudelman; see note 20.

2. *Narrative Consequences of the Novum*

2.1 The presence of the novum as the determining factor of an SF narration is crucially testable in its explanatory power for the basic narrative strategies in this genre. First of all, the dominance or hegemony of the cognitive novelty means that an SF narration is not only a tale that includes this or that SF element or aspect: utopian strivings or dystopian terrors of some kind, as in the majority of world literatures; moral allegories or transcendental visions of other worlds, better or worse than our own, as in much literature down to Milton, Swedenborg, and countless imitators; use of new technological gadgets, as in many James Bond tales; etc. An SF narration is a fiction in which the SF element or aspect, the novum, is so central and significant that it determines the whole narrative logic (or at least the overriding narrative logic) regardless of any impurities that might be present.[10]

2.2 Furthermore, the novum intensifies and radicalizes that movement across the boundary of a semantic field (defined by the author's cultural norm) which always constitutes the fictional event.[11] In "naturalistic" fiction this boundary is iconic and isomorphic: the transgression of the cultural, of an ontological norm: by an ontic change in the character/ Bovary's adultery stands for an adultery. In SF, or at least in its determining events, it is not iconic but allomorphic: a transgression of the cultural norm is signified by the transgression of a more than merely cultural, or an ontological norm: by an ontic change in the character/ agent's reality either because of his displacement in space and/or time, or because the reality itself changes around him. For lack of a better term, I can only say that the novelty makes for the SF narration's specific *ontolytic* effect and properties. Unlike the fantasy tale or the mythological tale, SF does not posit another superordinated and "more real" reality but an

[10]A major objection against so-called "thematic" studies of SF elements and aspects, from James O. Bailey's *Pilgrims Through Time and Space* (New York: Argus Books, 1947)—in its time no doubt a pioneering work—to present-day atomistic and positivistic SF critics, is that these studies ignore the determining feature of what they are studying: the narrative logic of a fictional tale. Correlatively, they tend to become boring catalogs of raisins picked out of the narrative cake, and shrivelled in the process. This does not mean that critical discussions of, e.g., artificial satellites, biological mutations, or new sexual mores in SF (or other fiction) cannot be, for some strictly limited purposes, found useful; and for such purposes we should probably know where the mutations, satellites, or sex patterns first appeared and how they spread. But we should not be lured by this very peripheral necessity into annexing any and every tale with a new gadget or psychic procedure into SF, as Bailey did with Wilkie Collins' *Moonstone* and Thomas Hardy's *Two on a Tower*. SF scholarship that does this is sawing off the branch on which it is sitting: for if these and such works are SF just like, say, Wells's *Invisible Man*, then in fact there is no such thing as SF.

[11]Jurij Lotman, *The Structure of the Artistic Text*, Michigan Slavic Contributions No. 7 (Ann Arbor, 1977), pp. 229 ff.

alternative on the same ontological level as the author's empirical reality. Therefore, the necessary correlate of the novum is an *alternate reality*, one that possesses a *different historical time* corresponding to different human relationships and sociocultural norms actualized by the narration. This new reality presupposes the existence of the author's empirical reality, since it can be gauged and understood only as the empirical reality modified in such-and-such ways. Though SF cannot by definition be an orthodox allegory with any one-to-one correspondence of its elements to elements in the author's reality, its specific modality of existence is a feedback oscillation that moves now from the author's and implied reader's norm of reality to the narratively actualized novum, in order to understand the plot-events, and now back from those novelties to the author's reality, in order to see it afresh from the new perspective gained. This oscillation, called estrangement by Shklovsky and Brecht, is no doubt a consequence of every poetic, dramatic, scientific, in brief *semantic* novum. However, in SF its second pole is a narrative reality sufficiently autonomous and intransitive to be explored at length as to its own properties and the human relationships it implies. (For though mutants or Martians, ants or intelligent nautiloids, can be used as signifiers, they can only signify imaginable human relationships, given that we can—at least so far—imagine no other ones.)

2.3 The oscillation between the author's "zero world" and the new reality induces the narrative necessity of a means of reality-displacement. As far as I can see, there are two such devices: a voyage to a new locus, and a *catalyzer* transforming the author's environment to a new locus; examples for the two could be Wells's *Time Machine* and *Invisible Man*. The first case is better suited to a simultaneous and the second to a gradual introduction of a new reality; no doubt, all kinds of contamination and twists on these two means are thinkable. When the *in medias res* technique is used in any particular SF tale, the means of displacement can not only be told in retrospective but even totally atrophy (more easily in a space/time displacement: our hero is simply a native of elsewhere/elsewhen). However, this semblance conceals the presence of displacement in a zero-form, usually as a convention tacitly extrapolated from earlier stories; the history of the genre is the missing link that made it possible, e.g., tales in another space/time without any textual reference to that of the author (as in most good SF novels of the last twenty years).

2.4 The concept of novum also illuminates the historical vicissitudes of justifying the reality-displacement. In naturalistic tales, the voyage can only start in the author's space, and the account of the new reality must eventually come back to that space. This leads to real problems in observing a naturalistic plausibility. It would then be logically necessary that the account of such a sensational voyage to a new reality in turn become a catalyzer, inducing changes in the author's and reader's environment. Since this in fact—as the reader knows—has not happened, naturalistic SF has had to invent a number of lightning-rods to dissipate such expectations. Verne pretended not to notice its necessity, while Wells in *War of the Worlds* pretended we all knew it already; these are ploys which make those novels sound as if they assumed an alternate time-stream in which Nemo or the Martians had in fact (as different from the reader's time-stream) been the scourge of the seas or of England. Many earlier writers went through other extraordinary contortions to satisfy naturalistic plausibility, usually a contamination of the "manuscript in a bottle" device (the news of a voyage to the Moon just arrived by volcano and served piping hot to you, dear reader) and the "lost invention" device (a one-shot novelty confined to the experience of a few people due to the loss of the invention) as in *The First Men on the Moon*. The most plausible variable for manipulating was *time*, since putting the tale into the future immediately dispensed the author from any account of empirical plausibility. The shift of SF from space into future time is due to an interaction of this narrative convenience, stunted within positivist ideology, with the strong tendency toward temporal extrapolation inherent in a capitalist economy, whose profits and *excelsior* ideals are always expected in a future clock-time.

Thus, space was a fully plausible locus for SF only before the capitalist way of life, from the very early tales about the happy or unhappy valley or island—known to almost all tribal and ancient societies—to More and Swift. An Earthly Paradise or Cockayne tale, a humanist dialog and satire, all happen in a literary or imaginative space not subject to positivistic plausibility. But a fully triumphant bourgeoisie introduces an epoch-making epistemological break into human imagination, by which linear or clock-time becomes the space of human development because it is the space of capitalist industrial production. Capital, the new historical shape of property, has in principle no limits in extrapolated time, unlike the spatial dominions of even the largest feudal landowner. Through a powerful system of mediations infusing the whole human existence, time finally becomes the equivalent of money and, thus, of all things. The positivist ideology followed capitalist practice in finally resolving time "into an exactly delimited, quantifiable continuum filled with quantifiable 'things' . . . in short, [time] becomes space."[12]

[12]Georg Lukács, *History and Class Consciousness* (London: Merlin Press, 1971), p. 90. The whole seminal essay "Reification and the Consciousness of the Proletariat," developing insights from Marx's *Capital*, is to be consulted; also Lewis Mumford, *Technics and Civilization* (New York: Harcourt Brace Jovanovich, Inc., 1963). I have tried to apply Lukács' ideas on quantification and reification in my essays "On Individualist World View in Drama," *Zagadnienia rodzajów literackih*, No. 1 (1966), and "Communication in Quantified Space," *Clio* (Oct. 1974) which deals with Verne as the bard of movement in such a quantified space.

Imaginative times and spaces are now resolved into "positive," quantified ones. All existential alternatives, for better or worse, shift into such a spatialized future, which now becomes the vast ocean on whose other shore the alternative island is to be situated. Positivism shunts SF into anticipation, a form more activistic than the spatial *exemplum* because it is achievable in the implied reader's own space. When the industrial revolution becomes divorced from the democratic one—a divorce which is the fundamental political event of the bourgeois epoch—activism becomes exasperated and leads to the demands for another epistemological break, signalled by Blake's Jerusalem in England's green and pleasant land and the cosmic "passionate attraction" of Fourier's phalansteries. Such imaginative energies converge into Marx, the great prefigurator of the imaginative shift still being consummated in our times. Rather than post-industrial (a fairly reified and vague term), I would tend to call the new *épistémè*—since it is marked by names such as Einstein, Picasso, Eisenstein, Brecht, and Marx—one of spatiotemporal covariance, simulsequentialism, or humanist relativism and estrangement: in brief, one of alternate historical realities. I would argue that in such a historical perspective, all significant SF from Zamiatin, Čapek, and Lem to Le Guin, Disch, and Delany is neither simply spatial, as in Lucian or More, nor simply temporal, as in all the followers of *The Time Machine* and *When the Sleeper Wakes,* but spatiotemporal. It is spatiotemporal in a number of very interesting ways, all of which approximate a reinvention and putting to new uses of the pre-capitalist and pre-individualistic analogic times and spaces of the human imagination.

The main difference with such medieval and pre-medieval conceptions could perhaps be expressed in terms of Destiny. As Lotman remarks, literary functions can be divided into two groups, the active forces and the obstacles.[13] Right down to Swift (in SF and in literature in general), the obstacles are inhuman and superhuman forces, at best to be ethically questioned by the tragic poet and hero but not to be materially influenced. Whether they are called gods, God, Destiny, or Nature is relatively less important than the fact that they are transcendental, empirically unchangeable. The great enlightening deed of the bourgeoisie was to reduce the universe to individuals and identify the obstacles too with men, who are reachable and perhaps removable by other individual men. I would imagine that a truly modern literature (and SF) corresponding to our epoch and its *épistémè* would be the third dialectical term to follow fatalistic collectivism and humanistic individualism. We have learned that the institutional and imaginative products of men—states, corporations, religions, wars, etc.—can very well become a Destiny for

[13]Lotman, p. 239.

each of us: tragedy is again possible in the twentieth century (the October Revolution and Second World War, Dubček and Allende *docent*), though it is the tragedy of blindness, of failed historical possibilities, rather than of lucidity. The obstacles are super-individual but not inhuman; they have the grandeur of the ancient Destiny but they can be overcome by other men banding together for the purpose. Men are the historical Destiny of man; the synthesis in this historical triad is a humanistic collectivism.

2.5 The alternate reality logically necessitated by and proceeding from the narrative kernel of the novum can only function in the oscillating feedback with the author's reality suggested earlier, because it is as a whole—or because some of its focal relationships are—an *analogy* to that empirical reality. However fantastic (in the sense of empirically unverifiable) the characters or worlds described, always *de nobis fabula narratur.* Though SF is not orthodox allegory, it transmits aesthetic information in direct proportion to its relevance and aesthetic quality. The alternative is for it to operate in semantic emptiness spiced with melodramatic sensationalism as a compensatory satisfaction, in a runaway feedback system with corrupt audience taste instead of with cognition of tendencies in the social practice of human relationships.[14] The clear dominance of that Kitsch alternative in the present historical period should not, however, prevent us from discussing the significant models of SF, its horizons and yardsticks.

Ten years ago I proposed to define the heuristic models of SF as extrapolation and analogy. Extrapolation starts from a cognitive hypothesis incarnated in the nucleus of the fable and directly extrapolates it into the future; in analogy, cognition derives only from the final import or message of the tale, and may perhaps be only indirectly applicable to pressing problems in the author's environment. This fairly frequently quoted analysis was, I hope, useful in combating the then unchallenged definition of SF as extrapolation, but is seems insufficient today. In that essay I noted that any futurological function SF might have was strictly secondary, and stressing it was dangerous since it tended to press upon SF the role of a popularizer of the reigning ideology of the day (techno-

[14]"The information gained, concerning a hypothesis, may perhaps be thought of as the ratio of the a posteriori to the a priori probabilities (strictly the logarithm of this ratio)"; see Colin Cherry, *On Human Communication* (Cambridge, Mass.: MIT Press, 1966), p. 63. Thus, the information gained from a work of literature is a logarithmic (i.e., alas, much diminished) ratio of the existential possibilities imaginable and understandable by an ideal reader after reading to those imaginable and understandable before the reading. The information is a function of the rearrangement of the reader's understanding of human relationships. "In general, where we speak of information we should use the word form," argues René Thom in his impressive *Stabilité structurelle et morphogénèse* (Reading, Mass., 1972), p. 133.

cratic, psionic, utopian, dystopian, hip, or whatever). My skepticism toward both futurology and extrapolation as an aesthetic possibility has grown considerably. Therefore, I believe that the trouble with my original division was not that, as some have thought,[15] it contained too few but rather too many models. The defining of SF as extrapolation (to which the title of a critical journal devoted to SF still witnesses) should be decently and deeply buried. It seems quite clear by now that SF is material for futurology (if at all) only in the very restricted sense of reflecting on the author's own historical period and the possibilities inherent in it: *1984* is about collective human relationships in 1948 as certain aspects of or elements within Orwell's mind saw it, and *2001* about collective human relationships in 1967 as certain aspects of or elements within Kubrick's mind saw it. Thus, any significant SF text is always to be read as an analogy, somewhere between a vague symbol and a precisely aimed parable; and extrapolative SF was only a historical delusion of technocratic ideology, ca. 1920–1950.

Indeed, extrapolation itself as a scientific procedure (and not pure arithmetic formalization) is predicated upon a strict, or if you wish crude, analogy between the points from and to which the extrapolation is carried out: extrapolation is a one-dimensional, scientistic limit-case of analogy. As Peirce put it, a scientific "effect [or "phenomenon"] consists in the fact that when an experimentalist shall come to *act* ACCORDING TO A CERTAIN SCHEME THAT HE HAS IN MIND then will something else happen, and shatter the doubts of sceptics, like the celestial fire upon the altar of Elijah."[16] Specifically, the SF "future-story" has been well identified by Raymond Williams as

> the finding and materialization of a *formula* about society. A particular pattern is abstracted, from the sum of social experience, and a society is created from this pattern. . . . The "future" device (usually only a device, for nearly always it is obviously contemporary society that is being written about . . .) removes the ordinary tension between the selected pattern and normal observation.[17]

Clearly, neither is the future a quantitatively measurable space, nor will the ensemble of human relationships stand still for one or more generations in order for a single element or group of elements to be

[15]See Philmus, note 4, and Fredric Jameson, "Generic Discontinuities in SF" and "World Reduction in Le Guin," both in *Science-Fiction Studies,* ed. R. D. Mullen and Darko Suvin, pp. 28–39 and 251–60.

[16]Charles Sanders Peirce, "What Pragmatism Is," *Collected Papers,* ed. Charles Hartshorne and Paul Weiss (Cambridge, Mass.: Harvard Univ. Press, 1934), para. 425, p. 284. Capitals, but not italics, mine.

[17]Raymond Williams, *The Long Revolution* (Harmondsworth: Penguin Books, 1971), p. 307.

extrapolated against an unchanging background; this is the common invalidating premise of futurological as well as of openly fictional extrapolation. The future is always constituted both by a multiple criss-crossing of developments and—in human affairs—by intentions, desires, and beliefs rather than by quantifiable facts. It is Peirce's scheme or Williams' pattern rather than the end-point of a line.

The problem in constructing useful models for SF is, then, one of differentiating *within* analogy. If every SF tale is some kind of analogy—and I think that *The Time Machine* or *The Iron Heel,* Heinlein's *Future History* or Pohl-Kornbluth's *Space Merchants,* even Stapledon's *Last and First Men* or Yefremov's *Andromeda* are primarily fairly clear analogies to processes incubating in their author's epoch—then just what is in each case the degree and the kind of its anamorphic distortion, its "version" of reality? How is their implied reader supposed to respond to and deal with a narrative reality that is an inverted, reverted, converted, everted, averted, subverted Other to his certainties of Self and Norm—certainties which, as Hegel says, are clouded by their very illusion of evidence and proximity, *bekannt* but not *erkannt?*[18] A halfway illuminating answer to this group of questions would also clear up why some of these versions pretend—sometimes with conviction, most often by pure convention—to be situated in an extrapolated future.

2.6 A final narrative consequence of the novum would be that it shapes the SF "chronotope" (or chronotopes?). A chronotope is "the essential connection of temporal and spatial relationships, as shaped in literary art." In it, "the characteristics of time are unfolded in space, while space is given meaning and measured by time,"[19] and both are blended into a particular plot-structure. Now the novelty in SF can be either a new locus, or an agent (character) with new powers transforming the old locus, or a blend of both. The connection between the active forces (the protagonist[s]) and the obstacles to be reduced (the locus) determines the homogeneity of a tale. If the protagonists and the loci necessarily imply and richly reinforce each other—as do Wells's Time Traveller and the sequence of his devolutionary visions of the future, or Le Guin's Shevek, his physics, and the binary planetary sociopolitics and psychology of *The Dispossessed*—then we have a tale of a higher quality than the wish-dreams of, say, a Van Vogt, where all the obstacles are fake since the protagonist is a superman enforcing his will both on enemies and supposed allies.

As for plot-structures, if SF is organized around an irreversible and

[18]Georg Wilhelm Friedrich Hegel, *Phänomenologie des Geistes. Sämtliche Werke* (Leipzig: F. Meiner, 1949), II, p. 28.

[19]Mikhail Bakhtin, *Voprosy literatury i èstetiki* (Moscow, 1975), pp. 234–35.

significant change in its world and agents, then a simple addition of adventures, where *plus ça change plus c'est la même chose,* is an abuse of SF for purposes of trivial sensationalism: it degrades the genre to a simpler and less organized plot-structure. Nudelman has brilliantly demonstrated the incompatibility between the plot-structure of the cyclical detective tale, whose conclusion returns the universe "to its equilibrium and order"; the linear structures of the additive adventure tale; and the spiral structures of valid SF, whose plot alters the universe of the tale.[20] On the contrary, the easiest narrative way of driving a significant change home is to have the hero grow into it or better, to have the hero define it for the reader by growing with it. Much valid SF uses the plot-structure of the "educational novel" with its initially naive protagonist who by degrees arrives at some understanding of the novum for himself and for the readers.

As these two examples and other discussions in this essay may indicate, it should be possible to engage in analytic evaluation of SF that would be neither purely ideological nor purely formalistic by starting with the necessities of literary structure brought about by some variant of a novum.

3. *The Novum and History*

3.1 The novum as a creative, and especially as an aesthetic, category is not to be fully or even centrally explained by formal aspects such as innovation, surprise, reshaping, or estrangement, important and indispensable though these aspects or factors are.[21] The new is always a *historical* category since it is always determined by historical forces which both bring it about in social practice (including art) and make for new semantic meanings that crystallize the novum in human consciousnesses. An analysis of SF is necessarily faced with the question of why and how the newness was recognizable as newness at the moment it appeared, of what ways of understanding, horizons, and interests were implicit in the novum and required for it. The novelty is related, not merely as direct reflection but also as complex prefiguration or negation, to new historical forces and patterns, in the final instance, to new possibilities of human relationships. An aesthetic novum is either a translation of historical cognition and ethics *into* form, or (in our

[20]Rafail Nudelman, "An Approach to the Structure of Le Guin's SF," in *Science-Fiction Studies,* ed. R. D. Mullen and Darko Suvin, pp. 240–50.

[21]See, for development of estrangement and similar notions after the Formalists and Brecht, Hans Robert Jauss, *Literaturgeschichte als Provokation* (Frankfurt am Main: Suhrkamp, 1970), as well as critiques of and improvements on Jauss handily assembled in *Sozialgeschichte und Wirkungsästhetik,* ed. Peter Uwe Hohendahl (Frankfurt am Main: Athenäum-Fischer-Taschenbuch-Verlag, 1974).

age perhaps more often) a creation of historical cognition and ethics *as* form.

3.2 Probably the most important consequence of understanding SF as a symbolic system centered on a novum (which must be cognitively validated within the narrative reality of the tale and its interaction with reader expectations) is that the novum has to be convincingly explained in concrete even if imaginary terms, i.e., in terms of the *specific* time, place, agents, and cosmic and social totality of each tale. This means that, in principle, SF has to be judged, like naturalistic or "realistic" fiction and quite unlike the supernatural genres, by the density and richness of objects and agents described in the microcosm of the text. We could set up a further Hegelian triad where the thesis would be naturalistic fiction, which has an empirically validated effect of reality, the antithesis would be supernatural genres, which lack such an effect, and the synthesis would be SF, in which the effect of reality is validated by a cognitive innovation. Obversely, the particular essential novum of an SF tale has to be judged by how much new insight into imaginary but coherent and this-worldly, or *historical,* relationships it affords.

3.3 In view of this doubly historical character of the SF novum, born in history and judged in it, the novum has to be differentiated not only according to its degree of magnitude and of cognitive validation, but also according to its *degree of relevance*. What is *possible* should be differentiated not only from what is already real but also from what is equally empirically unreal but *necessary*. Not all possible novelties will be equally relevant, or of equally lasting relevance, from the point of view of, first, human development, and second, a positive human development. Obviously, this categorization implies, first, that there are some law-like tendencies in men's social and cosmic history, and second, that today (if we are intelligent and lucky enough) we can judge these tendencies as parts of a spectrum that runs from positive to negative. I subscribe to both of these propositions, and will not argue them here—mainly because I cannot think of any halfway significant SF narration which does not in some way subscribe to them.

A novum can be both superficially sweeping and cognitively validated as not impossible, and yet of very limited or brief relevance; if so, its relationship to a really relevant novum will be the same as the relationship of the yearly pseudo-novelty of "new and improved" and "revolutionary" car-types or clothing fashions to a really radical novelty: say, a social revolution or change in scientific paradigms. The pseudo-novum's vitality will not be that of a tree, an animal species, or a belief but, to quote Bergson, the explosive, spurting "whoosh" of a howitzer shell exploding into successively smaller fragments, or "of an immense fireworks, which continually emits further firesparks from its midst."[22] In brief, a novum is an instantly obsolescent fake unless it in some way participates in what Bloch called the "front-line of historical process"—which for him and for me as Marxists means a process intimately concerned with strivings for a de-alienation of men and their social life. Capricious contingencies, consequent upon the market competition and tied to copyright and patent law, have a built-in limit and taboo defined precisely by the untouchable sanctity of competition—so palpable an ideology in much SF. Capricious contingencies, of brief and narrow relevance, particular rather than general, make for a superficial change rather than for a true novelty which makes for human relationships so qualitatively different from those dominant in the author's reality that they cannot be translated back merely by a change of costume. All space operas can be translated back into the Social Darwinism of the Westerns and similar adventure-tales by substituting colts for ray-guns and Indians for the slimy monsters of Betelgeuse. Most novels by Asimov can be returned to their detective-story model by a somewhat more complex system of substitutions, by which, e.g., the *Second Foundation* came from Poe's *Purloined Letter*.

3.4 Since freedom is the possibility of something new and truly different coming about, "the possibility of making it different,"[23] the distinction between a true and a fake novum is not only a key to aesthetic quality in SF but also to its ethico-political liberating qualities. As always in art, ethical pathos and effect or communal (political) relevance are the obverse of aesthetic consistency. Both of them fuse in the realization that, finally, the only consistent novelty is one that constitutes an open-ended system "which possesses its novum continually both in itself and before itself; as befits the unfinished state of the world, nowhere determined by any transcendental supra-worldly formula."[24] This connects with my argument about validation for SF being based on science as an open-ended corpus of knowledge; the argument can now be seen to be ultimately and solidly anchored to the bedrock fact that there is no end to history, and that we and our ideologies are not the end-product history has been laboring for from the time of the first sabre-toothed tigers and Meso-

[22]Henri Bergson, *L'évolution créatrice* (Paris: F. Alcan, 1907), p. 99 and 270; see also Bloch's comment on him in *Das Prinzip Hoffnung*, p. 231. My whole argument in 3.1-3.4 is fundamentally indebted to Bloch. For more on originality within a capitalist market see Bertolt Brecht, *Gesammelte Werke* (Frankfurt am Main: Suhrkamp, 1973), 1–20 passim-e.g., 15, pp. 199–200; and Theodor W. Adorno, *Aesthetische Theorie* (Frankfurt am Main: Suhrkamp, 1970), pp. 257ff.

[23]Bloch, *Experimentum Mundi*, p. 139; see also Antonio Gramsci, *Il Materialismo storico e la filosofia di Benedetto Croce* (Torino: 1948), quoted from *Selections from the Prison Notebooks*, ed. Quintin Hoare and Geoffrey Nowell-Smith (New York: International Publishers, 1971), p. 360, to which I am also much indebted.

[24]Bloch, *op. cit*, p. 143.

potamian city-states. It follows that SF will be more significant and truly relevant the more clearly it eschews final solutions, be they the static utopia of the Plato-Moore type, the more fashionable static dystopia of the Huxley-Orwell type, or any similar metamorphosis of the Apocalypse (let us remember that the end of time in the Apocalypse encompasses not only the ultimate chaos but also the ultimate divine order).[25]

3.5 An imaginary history, each time to be reimagined afresh in its human significance and values, may perhaps borrow some narrative patterns from mythological tales, but the "novelty" of gods validated by unexplained supersciences at the beck and call of the Cambridge School's or von Däniken's super-mortals is a pseudo-novelty. SF's analogical historicity may or may not be mythomorphic, but it cannot be mythopoetic in any sense except the most trivial one of possessing "a vast sweep" or "a sense of wonder"; it is another superannuated slogan of much SF criticism due for a deserved retirement into the same limbo as extrapolation. For myth is re-enactment, eternal return, and the opposite of a creative human freedom.

An objection might be raised with reference to the same bookstore shelves mentioned earlier, that after one subtracts the more or less supernatural tales (science-fantasy, sword-and-sorcery, etc.) from what is usually sold as SF, ninety percent of what remains will have plot-structures escaping from history into Westerns, additive sensationalist adventures, or rehashes of mythography. However, I take my stand with Kant's dictum that a thousand years of any given state of affairs do not make that state necessarily right. Rather, reasons for the wrongness should be sought.

3.6 This analysis has finally reached the point where history, in the guise of analogical historicity, is found to be the next and crucial step in the understanding of SF: story is always also history, and SF is always also a certain type of imaginative historical tale (which it would be useful to compare to the historical novel).[26] All the epistemological, ideological, and narrative implications and correlatives of the novum I have discussed lead to the conclusion that significant SF is in fact a roundabout way of commenting on the author's collective context, often resulting in a surprisingly concrete and sharp-sighted comment at that. Even where

[25] I have attempted to expand on this in my "The Open-Ended Parables of Stanislaw Lem and *Solaris*," afterword by Stanislaw Lem, *Solaris* (New York: Berkley Publishing Corp., 1976); it is incorporated into a parallel to the orthodox Soviet and American SF models in my "Stanislaw Lem und das mitteleuropäische soziale Bewusstsein der Science Fiction" in *Insel Almanach auf das Jahr 1976: Stanislaw Lem,* ed. Werner Berthel (Frankfurt am Main: Insel-Verlag, 1976).

[26] I am indebted for this idea, and for other seminal hints, to conversations with and unpublished manuscripts by Professor Fredric Jameson.

SF, sometimes strongly, suggests a flight from that context, it is an optical illusion and an epistemological trick. In all such SF that is significant, the escape is to a better vantage point from which to comprehend the human relations around the author. It is an escape from constrictive old norms into a different and alternative time-stream, a device for historical estrangement, and at least initially a readiness for new norms of reality, for the novum of de-alienating human history. In order to understand it properly, I believe, the critic will have to integrate socio-historical knowledge into formal knowledge, diachrony into synchrony. History has not ended with the "post-industrial" society: as Bloch said, Judgment Day is also Genesis, and Genesis is every day.

Signs of W$^{a}_{o}$nder

TERESA DE LAURETIS

PRO/LOGUE IN THE FUTURE

Toward the end of the twentieth century, in a decentered region of Sol's Terra which we now call the Arctic Riviera, two indigenous clans, known as Humanists and Technologists, met at a ritual gathering for the purpose of solving what they called ideological differences. This archeological curiosity would have no interest for us were it not for the fact that a few of the Humanist clan, beings endowed with a vision still very rare in those days, met separately and discussed signmaking. Of course they had another name for it, a strange word-sound—"SF". Their culture was still very primitive and they naively thought that "SF" was just one among many forms of social creativity. Nonetheless, the following text, which according to the legend was produced during that ritual gathering, can be read as an interesting historical sign.

(from *Galactic Fictions*)

PRE/TEXT

> Because it was clear that, independent of signs, space didn't exist and perhaps had never existed.
>
> (Italo Calvino, *Cosmicomics*)

In its relatively short life span, SF has already acquired a place in literary history marked by a genealogy of influence, a generic topology, a typology of motifs, a sociology of, etcetera. It also has a literary history of its own, complete with a classical period, a New Wave, even, some claim, a postmodern phase. But for the purposes of this essay *on* SF, let us step out of the established categories of literary history and, suspending disbelief, enter the magnetic field of play that is the language *of* SF. We will then assume that, in a hypothetical universe of critical discourse, one can define SF as a mode of writing and a manner of reading, a way of telling what has not happened and of seeing what is not apparent. We will further assume that both the hypothetical universe of critical discourse and the science fictional universe on which the former speculates are semiotic fictions. And that, independent of signs, neither universe would exist.

To posit SF as a mode of writing, a manner of reading, is not to disregard the fact that SF is inscribed in narrative forms but to explore the textual processes that coexist with narrativization and counter its tendency to totalize meaning. In this hypothetical critical perspective, the text is viewed neither as an entity or an object, nor a system of relations, a structural hierarchy of levels, an originary locus of meaning; it is not/does not have a unit or a single principle of construction. But, on the other hand, neither is the text viewed formalistically as an endless play of differences, deferments or dislocations internal to, or within the text itself. On the contrary, the science fictional text appears as a discontinuous set of sign-functions which produces an indefinite set of semantic constructs by dislocating the subject in the reading process, and requiring exits from, as well as providing entrances into, the text. Thus SF as a way of writing and reading is dependent upon what could be called the sign-work: a partically structured movement in and out of the text, establishing temporary associations and provisional encounters of signs and meanings (sign-functions) which, by constantly shifting with the context, in the reading process, do not resolve in a totalization of meaning but in fact resist the strong pull toward narrative closure and disrupt the mechanisms of narrativity.[1]

[1]The term "sign-work" is coined by analogy on Freud's "dream-work" (*Traumarbeit*). The notion of "sign-function" in Umberto Eco, *A Theory of Semiotics* (Bloomington: Indiana University Press, 1976) is more flexible than the traditional "sign," which is usually

Since the notion of sign-work is the only instrument of analysis available to us in this hypothetical critical discourse, I must first define my use of the term "sign" and then explain how I see the work of signs in SF. Signs are not things, but complex functions: like texts, they are constituted of multiply layered webs of relationships and associations. Every sign is the close encounter of an emitter and a receiver; and every sign is the meeting ground of a material/physical sign-vehicle (sound, image, object) with a galaxy of semantic constructs (meaning, mental image, cultural unit).[2] The modes of production and circulation of signs, and therefore meanings, are culture specific: they are not given once and for all in the Book of Genesis or in the Oxford English Dictionary but constantly change or shift with the historical context. Unlike myths (which, by the way, are *not* ageless), SF and its signs belong in advanced technological societies.

Signs, cognition, and social practice

I think SF has a way of using signs that is potentially creative of new forms of social imagination; creative in the sense of mapping out areas where cultural change *could* take place, of envisioning a different order of relationships between people and between people and things, a different conceptualization of social existence, inclusive of physical and material existence. I am not suggesting that SF is a "revolutionary" art form in the simplistic sense of the term. Clearly, art forms by themselves do not change things; only people can change "things," i.e., their social relations, which include the relations of humans to the physical world. But our knowledge of material and social reality, as well as the possibility to change it, is the result of semiosic processes. In other words, cognition, perception, conceptualization involve the continuous activity of producing and exchanging signs and meanings. Thus what we know, what our reality is for us, is determined by the social production of signs and meanings and by its purposes, that is to say, by social practice. Signs are both the manifestations and the agents of social practice.[3]

identified with the linguistic sign or a single unit of a given system, whereas "sign-function" includes various types of significant units such as strings of signs, texts, and macrotexts. A sign-function is "an element of an *expression plane* conventionally correlated to one (or several) elements of a *content plane*" (p. 48). The term "sign" continues to be used by Eco, and is used in this article, for reasons of brevity, but always in the sense of "sign-function" defined above. For the term "narrativization," indicating the processes, structure, and movement of narrative, see Stephen Heath, "Narrative Space," *Screen*, 17 (Autumn 1976), 68–112.

[2]With the term "sign-vehicle," preferable to the Saussurian "signifier" (*signifiant*), Eco indicates the physical component of a sign-function. See *A Theory of Semiotics*, pp. 48–49 and ff.

[3]See Louis J. Prieto, *Pertinence et pratique: Essais de sémiologie* (Paris: Minuit, 1975),

If SF signs (which are still to be defined) point to other possible worlds than the one we know, how do they do so? Why or how are they signs of wonder? Samuel Delany describes SF as that which has not happened, while fantasy is what cannot happen.[4] (How do we *know* what can or cannot happen?) Do we really wonder at what cannot happen, or only at what *could* happen? Traditionally there have been utopias and dystopias, as visions of what people hoped or feared could happen. Very recently the fictions have become ambiguous, puzzling, unfamiliar maps of heterotopias.[5]

> The great Khan owns an atlas in which are gathered the maps of all the cities: those whose walls rest on solid foundations, those which fell in ruins and were swallowed up by the sand, those that will exist one day and in whose place now only hares' holes gape. . . . The atlas has these qualities: it reveals the form of cities that do not yet have a form or a name. . . . The catalogue of forms is endless: until every shape has found its city, new cities will continue to be born.[6]

---

especially Chapter 5, pp. 143–65.

[4]Samuel R. Delany, "About Five Thousand Seven Hundred and Fifty Words," in *The Jewel-Hinged Jaw* (Elisabethtown, New York: Dragon Press, 1977), pp. 43–44.

[5]Delany's novel *Triton* (New York: Bantam, 1976) is explicitly subtitled "An Ambiguous Heterotopia." The author plays on the several meanings of the word: "displacement in or difference of position: as a) deviation of an organ from the normal position, b) an abnormal habitat, c) the grafting of tissue into an abnormal location" (*Webster Third New International Dictionary,* 1966). The dissolution of meaning and of the subject's unity in language is also the effect of heterotopias, according to Michel Foucault. In the Preface to *The Order of Things* (New York: Pantheon Books, 1970), originally *Les Mots et les choses* (Paris: Gallimard, 1966), he states: "in such a state, things are 'laid,' 'placed,' 'arranged' in sites so different from one another that it is impossible to find a place of residence for them, to define a *common locus* beneath them all. *Utopias* afford consolation: although they have no real locality there is nevertheless a fantastic, untroubled region in which they are able to unfold. . . . *Heterotopias* are disturbing, probably because they secretly undermine language, because they make it impossible to name this *and* that, because they shatter or tangle common names, because they destroy 'syntax' in advance, and not only the syntax with which we construct sentences but also that less apparent syntax which causes words and things (next to and also opposite one another) to 'hold together'" (pp. xvii-xviii). In 1969, describing how "modern s-f has gone beyond this irreconcilable Utopian/Dystopian conflict to produce a more fruitful model against which to compare human development," Delany stated: "Now the writers began to explore these infinitely multiplicated worlds, filled with wondrous things, where the roads and the paintings moved . . . where travel could be instantaneous between anywhere and anywhere else, where the sky was metal, and women wore live goldfish in the transparent heels of their shoes. Within these worlds, the impossible relieves the probable, and the possible illuminates the improbable. And the author's aim is neither to condemn nor to condone, but to explore both the worlds and their behaviors for the sake of exploration, again an aim far closer to poetry than to any sociological brand of fiction" (*The Jewel-Hinged Jaw,* pp. 144–45).

[6]Italo Calvino, *Invisible Cities* (New York: Harcourt Brace Jovanovich, 1974), pp. 137–39. Calvino is better known in the United States as a science fiction writer for his *Cosmicomics*

Both Joanna Russ and Samuel Delany have stated in their critical writings, and shown in their practice of fiction, that SF uses language and narrative signs in a literal way.[7] A phrase such as "the door dilated" (cited by Delany from Harlan Ellison from Robert Heinlein) which in poetry or in non-SF would be read/could be read only as a metaphor, in SF is (to be) read literally, and therefore states the possibility that other physical and/or perceptual laws exist in the universe than the ones we have been taught or believe exist.

Another example is the ocean of Lem's *Solaris.*[8] Our reading protocol, derived from a long-established metaphysical and anthropocentric tradition, keeps urging us to read that ocean as a teleological force, a Will, a purposeful and intelligent activity. But the science-fictional sign, by generating multiple possibilities of reading against its narrative context, forces us to question our reading protocol first, and then hopefully our assumptions about "reality." Another sign of this kind is *Babel-17:*[9] its meaning, not given from the start but to be discovered in the process of reading, is a multilevel web of forces, relationships, and events that cannot be read as simple metaphors for existing realities but point forward to a possible world that "has not happened."

Why is it significant that these signs of SF are taken as literal statements and not as metaphors? A metaphor is an ambivalent sign, a word or expression which suggests that several common and different characteristics are shared by two distinct referents: e.g., /a masculine woman/ means "a woman who in addition to certain characteristics of her sex also displays traits that belong to another, entirely different sex." This, of course, is possible only if the concept "man" and the concept "woman" refer to two entirely distinct and in fact opposable/opposite sets or galaxies of human traits. So metaphor joins two sets of semantic units, but at the same time confirms their discreteness. Metaphor seems to have the power to change the world (as Rimbaud said and Lévi-Strauss repeated after him)[10] but in fact merely establishes itself as autonomous conceptual process in the sphere of language, separated from, and *above,* the order of material reality. The sign-as-metaphor can only refer to

(New York: Harcourt, 1968) and *t zero* (New York: Harcourt, 1969), which in Italy, due to its different cultural context, are read as literature. Like Bellona in Delany's *Dhalgren* (New York: Bantam, 1975), Calvino's *Cities* are maps of heterotopias, fictional intersections of technology and history.

[7]Joanna Russ, "Towards An Aesthetics of Science Fiction," in *Science Fiction Studies,* ed. R. D. Mullen and Darko Suvin (Boston: Gregg Press, 1976), pp. 8–15; Samuel R. Delany, *The Jewel-Hinged Jaw,* pp. 280–84.

[8]Stanislaw Lem, *Solaris* (New York: Berkley Medallion Books, 1971).

[9]Samuel R. Delany, *Babel-17* (New York: Ace Books, Inc., 1966).

[10]Claude Lévi-Strauss, *Structural Anthropology* (Garden City, New York: Doubleday, 1967), p. 197.

sign, can only slide over the world without ever coming into contact with it, perpetuating itself in endless difference.

Now, since signs are what our social existence is made of, it is important to conceptualize signs in a way that will allow us to use signs, rather than be used by them. (It is important to realize that signs do not merely "exist," in the ionosphere, but are produced by and circulate among social beings. The question is, of course, who produces them and in which manner they circulate—and this is a political question.) To re-literalize language has the effect of assigning to verbal signs some kind of use-value in addition to their exchange value, of going against the entropic enterprise of a society in which signs circulate bearing no relation to the material reality of the users, just as the wide and "free" circulation of money from hand to hand, from country to country, disguises the hard fact that it really always remains in the same hands.[11]

Silly as it seems to some people, the use of /Ms./ instead of /Miss/ or /Mrs./ has a real use-value: it tells women that they need not see themselves as someone's property—either the father's or the husband's. The definition of woman in terms of marriage, hence of belonging, of property, is something so old and so well internalized that we no longer perceive it as an active semantic marker. By creating the additional word /Ms./, women have shown that other social assumptions about women are possible and indeed necessary as well as desirable. This is what I suggest SF can do about many kinds of social relations involving class, (sub)culture, racial or sexual diversity. What cultural unit or galaxy of semantic traits corresponds to /female man/ as opposed to /masculine woman/ or /effeminate man/?[12] Obviously there are no given semantic traits for this new sign-vehicle in our present social reality. It can only be understood or become meaningful in a context/textus that posits other

[11]Cf. Jean Baudrillard, "Toward a Critique of the Political Economy of the Sign," *Sub-Stance*, 15 (1976), 115: "The rationality of the sign is based on the exclusion, the annihilation of all symbolic ambivalence in favor of a fixed and equational structure. The sign is a discriminator: it structures by exclusion. Henceforth crystallized on this exclusive structure, *designating* its fixed field, resigning all the rest and *assigning* the Sr and the Sd in a system of respective control, the sign presents itself as full value, positive, rational, exchangeable. All virtualities of meaning have been passed over the cutting edge of the structure. . . . Only ambivalence (to which we assign the strong meaning of rupture of value, beneath and beyond the sign-value and the emergence of the symbolic) puts into question readability, the false transparence of the sign, its use value (rational decoding) and its exchange value (the discourse of communication). *It puts an end to the poltical economy of the sign*, and hence to the respective definitions of the Sr and the Sd [and to] the process of signification in the classical sense." The concept of sign which Baudrillard justly criticizes is one narrowly derived from the Saussurian, linguistic definition and appropriated more or less analogically by other human sciences. Baudrillard's critique, therefore, does not apply to Eco's notion of the sign and in fact is cited here in support of my notion of the sign-work in SF.

[12]Joanna Russ, *The Female Man* (New York: Bantam, 1975).

assumptions about the world and other relations among people. And as we work our way through the text, associating, opposing, relating, remembering, or making unexpected discoveries, we are forced to break down patterns of expectation, recombine traits in different sets, revise our assumptions—we are forced to reconstruct time, space, and social relations. In other words, those sign-vehicles construct a new semantic universe, or re-organize the semantic space by reapportioning meaning in different ways and changing the value of words and actions.[13]

In this kind of fiction the adventure is not the protagonist's but the reader's. The relationship of SF with its readers is a very peculiar one, because it takes place not only within the SF text (as it does for all types of writing) but also outside the SF text proper, in the so-called fanzines. This latter and sociologically very interesting aspect of the writer-reader relationship cannot be explored here.[14] For the time being, I am concerned with the former aspect, the location—or better, the dislocation of the reader within the text, through the reading process. In the best of SF, the reader's sense of wOnder as awe, marvel, portent, revelation is replaced by a sense of wAndering through a mindscape both familiar and unfamiliar. Displaced from the central position of the knowledgeable observer, the reader stands on constantly shifting ground, on the margins of understanding, at the periphery of vision: hence the sense of wAnder, of being dislocated to another spacetime continuum where human

[13]The notion of "semantic space" is defined by Eco in *A Theory of Semiotics,* pp. 125–26. His discussion of creativity, especially the kind of creativity associated with art, which he calls invention (see pp. 245–76), is crucial to my argument here. Briefly, invention is "a mode of production whereby the producer of the sign-function chooses a new material continuum not yet segmented for that purpose and proposes a new way of organizing (of giving form to) it in order to *map* within it the formal pertinent element of content-type" (p. 245). Since, in this case, no previous convention exists, the sign producer must not only posit the correlation, but make it acceptable. By inventing both functives (expression and content) of the new correlation, and thus proposing a new convention, the artist can change the way in which culture sees the world. For a concise presentation of the fundamental concepts of Eco's *Theory,* see my "Semiosis Unlimited," *PTL: A Journal for Descriptive Poetics and Theory of Literature,* 2 (1977), 367–83.

[14]But I do wish to suggest that a study of fanzines (SF fan magazines) as historical texts and as evidence of the impact of SF on its readers could provide unusually rich material to reception theorists (cf. Peter Uwe Hohendahl, "Introduction to Reception Aesthetics," *New German Critique,* 10 [Winter 1977], 29–63). See, for example, Janice Bogstad's editorial, "The Science Fiction Connection: Readers and Writers in the SF Community," *Janus,* 10 (Winter 1977-78), 4–8; and the special issue of *Science-Fiction Studies,* 4 (November 1977) on "The Sociology of Science Fiction," in particular the articles by Linda Fleming, "The American SF Subculture," pp. 263–71, Charles Elkins, "An Approach to the Social Functions of American SF," pp. 228–32, and Albert I. Berger, "Science-Fiction Fans in Socio-Economic Perspective: Factors in the Social Consciousness of a Genre," pp. 232–46.

possibilities are discovered in the intersection of other signs with other meanings.[15]

> The Great Kahn's atlas contains also the maps of the promised lands visited in thought but not yet discovered or founded: New Atlantis, Utopia, the City of the Sun, Oceana, Tamoé, New Harmony, New Lanark, Icaria.
>
> Kublai asked Marco: "You, who go about exploring and who see signs, can tell me toward which of these futures the favoring winds are driving us."
>
> "For these ports I could not draw a route on the map or set a date for the landing. At times all I need is a brief glimpse, an opening in the midst of an incongruous landscape, a glint of lights in the fog, the dialogue of two passersby meeting in the crowd, and I think that, setting out from there, I will put together, piece by piece, the perfect city, made of fragments mixed with the rest, of instants separated by intervals, of signals one sends out, not knowing who receives them. If I tell you that the city toward which my journey tends is discontinuous in space and time, now scattered, now more condensed, you must not believe the search for it can stop. Perhaps while we speak, it is rising, scattered, within the confines of your empire; you can hunt for it, but only in the way I have said."[16]

The devil's advocate would argue, here, that much of the available body of literature is built, to some extent, in a similar way, on some device of defamiliarization (*ostranenie*). All poetry certainly is, and most non-realist fiction too. In "SF and the *Novum*" Darko Suvin responds that the distinctive feature of SF is "the narrative dominance or hegemony of a fictional novum validated by cognitive logic."[17] But my hypothetical universe of critical discourse does not include notions such as narrative dominance and unifying element. I have no critical language for speaking of a "novel," not to mention differentiating between an experimental novel and a detective story, or between a neo-gothic and a pornographic novel. I can, at best, speak of a metafictive mode or perhaps of certain modes of indeterminacy in fiction, like randomness or seriality, which can and often do co-exist more or less peacefully in the same text, even with the science fictional mode. As can many other things: plot, characters, themes, tropes, foreign words, songs, punctuation marks, graphics, misprints, ink smudges, pencil marks, marginal comments, remnants of torn pages, even occasionally a coffee stain. My discourse cannot analyze them, much less evaluate them. The sign-work is my only analytical tool here. For example, I could say that the sign-work of the

[15] Significantly, many of the fictions of science are also dislocated in/by SF signs: the anthropological fiction of field observation and scientific (objective) explanation of "primitive" cultures; the political fiction of bringing civilization through colonialism; or the sociological fiction of granting women equality by allowing them to behave like men.

[16] Calvino, *Invisible Cities,* p. 164.

[17] In this volume, p. 125.

metafictive mode is metaphoric, self-referential, intratextual, en-tropic; it cannot refer to anything outside the text for its semantic constructs are perched on the void at the center of Mind; it encloses the reader in the labyrinth of the text, in the solitaire pleasure of an endgame with no exit . . . and already I am sliding into the postmodern or postcognitive universes of paracriticism, critifiction, and so forth. So in the present essay, I can only attempt to describe the specificity of the science fictional mode using the analytical operational notion of sign-work.

If science and cognitive logic, as Suvin well says, "is the encompassing horizon of SF," technology is its diffuse landscape.[18] Technology is now, not only in a distant, science fictional future, an extension of our sensory capacities; it shapes our perception and cognitive processes, mediates our relationships with objects of the material and physical world, and our relationships with our own or other bodies. Technology is inseparable from labor, abstract knowledge, sexuality, art, daily activities, and fantasy. The fact that technology is utilized by multinational capitalism as an apparatus of ideological reproduction, designed to promote consensus by the production and distribution of information, does not mean that technology is outside social practice or can be removed from it. Technology is our historical context, political and personal.

> Both privileged and non-privileged thinkers are questioning our culture's context, scientific and otherwise, to an extent that makes trivial, by comparison, the blanket dismissal of all things with dials that glitter (or with latinate names in small print at the bottom of the labels) that the urban advocates of back-to-the-soil humanism sometimes claims to indulge. Within the city, because of the overdetermined context, even to attempt such a dismissal is simply to doom oneself to getting one's technology in grubbier packages, containing less efficient brands of it, and with the labels ripped off so that you cannot be sure what's inside. Those who actually *go* back to the soil are . . . concerned with exploring a folk technology, a very different process from "dismissal." And the radio-phonograph (solid-state circuitry) and the paperback book (computerized typesetting), just for examples, were integral parts of the exploration. That science fiction is the most popular literature in such places doesn't surprise.[19]

SF is unique among fictional modes in its capacity to deal with this

[18]And in this landscape both writer and reader inevitably move, in the manner of the semiotician whose approach, Eco says, "will not be like exploring the sea, where a ship's wake disappears as soon as it has passed, but more like exploring a forest where cart-trails or footprints do modify the explored landscape, so that the description the explorer gives of it must also take into account the ecological variations that he has produced" (*A Theory of Semiotics*, p. 29).

[19]Delany, *The Jewel-Hinged Jaw*, pp. 98–99.

historical reality of our age, a cultural reality which is shaped by technology in all its aspects—economic and psychological, public and private.

> Though few science fiction writers enjoy admitting it, much science fiction, especially of the nuts and bolts variety, reflects the major failure of the scientific context in which most technology presently occurs: the failure, in a world where specialization is a highly productive and valued commodity, to integrate its specialized products in any ecologically reasonable way. . . . What the urban humanist refuses to realize (and what the rural humanist often has no way of realizing) is that our culture's scientific context, which has given us the plow, the tape-recorder, insecticides, the butter-churn, and the bomb, is currently under an internal and informed onslaught as radical as our social context is suffering before the evidence of Women's Liberation, Gay Activism, Radical Psychiatry, or Black Power. Much science fiction inadvertently reflects the context's failure. The best science fiction explores the attack.[20]

SF's insistence on the materiality of the physical and technological environment of a possible "world," and its efforts to imagine what individual responses, what social relations obtain in such material environment, is what its literalization of language is all about.[21] It is the concrete, sensible specificity of its signs that allows SF language to be poetic rather than merely metaphoric, and re-charges the worn-out narrative patterns by posing a new relation between character and landscape or, in Delany's words, subverting the relative positions of figure

[20] *Ibid.*

[21] "Science fiction *is* science fiction because various bits of technological discourse (real, speculative, or pseudo)—that is to say the 'science'—are used to redeem various other sentences from the merely metaphorical. . . . Sometimes, as with the sentence 'The door dilated,' from Heinlein's *Beyond This Horizon*, the technological discourse that redeems it—in this case, discourse on the engineering of large-size iris apertures; and the sociological discourse on what such a technology would suggest about the entire culture—is not explicit in the text. . . . In other cases, such as the sentences from Bester's *The Stars My Destination* . . . the technological discourse that redeems them for the denotative description/presentation of incident *is* explicit in the text: "Sensation came to him, but filtered through a nervous system twisted and short-circuited by the PryE explosion. He was suffering from Synaesthesia, that rare condition in which perception receives messages from the objective world and relays these messages to the brain, but there in the brain the sensory perceptions are confused with one another.' In science fiction, 'science'—i.e., sentences displaying verbal emblems of scientific discourses—is used to literalize the meanings of other sentences for use in the construction of the fictional foreground. Such sentences as 'His world exploded,' or 'She turned on her left side,' as they subsume the proper technological discourse (of economics and cosmology in one; of switching circuitry and prosthetic surgery in the other), leave the banality of the emotionally muzzy metaphor, abandon the triviality of insomniac tossings, and through the labyrinth of technical possibility, become possible images of the impossible" (*The Jewel-Hinged Jaw*, pp. 92–93).

and ground.[22] Way back in 1953, an Italian critic, Sergio Solmi, said: "The most typical substance of SF" is not the novelistic plot or the moral-ideological allegory, but "the creation of an environment, the description of a machine, of a hypothetical human being, of imaginary cosmic structures. Imaginary, but materialized, like dreams that can be seen and touched."[23] What Solmi saw in SF was not futurologic anticipation or historicist extrapolation but "an anxious and absurdly precise effort to materialize abstract hypotheses and dreams," to make them concrete, sensible and experiential in the here and now.

In the context of my discourse, this would be restated in the following way: the "could happen" of SF is neither a past nor a future tense, but the present of another spatiotemporal order of discourse. The sign-work of SF, by re-literalizing language and giving it use-value, can oppose the entropy of social discourse and re-shape our semantic universe. Hence SF as a mode of writing and reading, as a textual and contextual production of signs and meanings, inscribes our cognitive and creative processes in what may be called the technological imagination. In tracing cognitive paths through the physical and material reality of the contemporary technological landscape and designing new maps of social reality, SF is perhaps the most innovative fictional mode of our historical creativity.

EPI/LOGUE IN TERRA

> The difficulty is not in grasping the idea that Greek art and epos are bound up with certain forms of social development. It lies rather in understanding why they still constitute for us a source of aesthetic enjoyment.
>
> (Karl Marx, *The Grundrisse*)

In every historical period, certain art forms (or certain literary forms, if we want to restrict our discourse to the art of writing alone) have become central to the episteme or historical vision of a given society: architecture and liturgic drama in the middle ages, when writing had no social impact; the invention of perspective in the Renaissance, which

[22]"There are very few 'ideas' in science fiction. The resonance between an idea and a landscape is what it's all about" (*The Jewel-Hinged Jaw*, p. 23). Also in Delany, "Of Sex, Objects, Signs, Systems, Sales, S-F, and Other Things," unpublished manuscript, p. 24: "Science fictional discourse redistributes the fictive charge between character and landscape in a manner different from mundane fiction. Science fiction makes the charge on the landscape much higher."

[23]Sergio Solmi, "Divagazioni sulla *Science-Fiction*, l'utopia e il tempo," *Nuovi Argomenti*, 5 (Novembre-Dicembre 1953), 20. This and all subsequent passages quoted are in my translation.

informed people's ways of seeing for centuries; the bourgeois and realist novel in the nineteenth century, cinema in this century, and so forth. If we compare it with traditional or postmodern fiction, we see that SF might, just might, be crucial from now on.

Most of the traditional forms of fiction, realist or romantic, like gothic novels, detective stories, biography or autobiography, even new journalism have actually outlived the social relations prevalent in the societies they represent and in which they developed as artistic forms. The "human condition" these fictions strive to "reflect," denounce or, less frequently, eulogize is simply not there.[24] Thus traditional or mundane fiction, today, is not a statement on "things as they are"; it is more likely a reflection on the ghost of Christmas past. On the other hand, much of the so-called experimental or postmodern fiction is a meditation on modernist literary techniques and on the anxieties of displacing the masters. (It has been said, I forget by whom, that all postmodern fiction is an attempt to rewrite *Finnegans Wake*.) While the overexposure of literary codes, the multiple reflections on the death of the novel, or the endless vigil of consciousness may have some appeal to the cultists of Art, the vision is contained within a cognitive horizon narrower than the present-day socio-historical imagination.[25]

The science fictional construction of a possible world, on the contrary, entails a conceptual reorganization of semantic space and therefore of material and social relations, and makes for an expanded cognitive horizon, an epic vision of our present social reality.

> The type of emotional charge that epic adventures arouse in the reader, by multiplying fantastically the number of possibilities and the potential for life, is still the same since the days of Homer, a manifestation of human passion and human restlessness. It's just that, today, the topography of the epos has expanded, its margins reaching out into the future.[26]

Thus Solmi, who adds:

> Ray Bradbury seems to assume the common places of science fiction, the machinery, the gadgets, in the same way in which Renaissance poets assumed in their epos the legends of the *chansons de geste:* as conventional

[24]And perhaps never was: as film studies have probed the formal construction of images and narrative patterns in neorealism, for example, the case has been made convincingly that neorealism was not at all a reflection or a photodynamic likeness of reality but rather a construction both of reality and of the viewer as ideological and perceptual subject.

[25]See the theoretical views of postmodernist writers expressed in Raymond Federman, ed., *Surfiction, Fiction now—and tomorrow* (Chicago: The Swallow Press, 1975), especially the introductory manifesto by Federman whose tone and proposed artistic program are remarkably similar to those of the Futurist manifestos over sixty years before.

[26]Solmi, p. 12.

material good for yielding tropes and narrative metaphors with which to express a contemporary and actual reality.[27]

In other words, fiction always refers to other fictions, whether abstract hypotheses of the past or dreams of the future. But all fictions are historically grounded in their present. SF speaks of the passion and the restlessness of the particular spacetime intersection that is our historical present. This is how Joanna Russ characterizes contemporary SF:

> Despite its ultra-American, individualistic muscle-flexing [American science fiction] is nonetheless collective in outlook, didactic, materialist. . . . Science fiction is the only modern literature to take work as its central and characteristic concern. . . . Like much "post-modern" literature (Nabokov, Borges) science fiction deals commonly, typically, and often insistently, with epistemology. . . . It is the only modern literature which attempts to assimilate imaginatively scientific knowledge about reality and the scientific method, as distinct from the merely practical changes science has made in our lives. . . . Science fiction is a positive response to the post-industrial world, not always in its content . . . but in its very assumptions, its very form.[28]

Having mentioned the words "epos" and "Homer," I must rush to clarify that my title is not a nostalgic plea for the wonder that was Greece. Far from it, my reference to the epic as a form of the social imagination, and my attributing to SF today the historical function that the epos had in classical Greece and in the European Renaissance, is meant to confront the question of the historicity of art forms. This question, far from being solved, is crucial to a theory of culture—the far horizon of any critical discourse, including the present one. Although I can offer no answer at the moment, I must at least attempt to reformulate the problem.

Referring to Marx's famous fragment, quoted at the beginning of this section, Karel Kosík states that Marx is not specifically investigating the Greek epos, but rather formulating a more general question: "How and why does a work of art outlast the conditions in which it originated?"[29] Kosík's answer is that works of art live on because "they have permanently enriched the human subject."[30]

[27] *Ibid,* p. 23

[28] "Towards An Aesthetic of Science Fiction," p. 13.

[29] Karel Kosik, "Historism and Historicism," *New German Critique,* 10 (Winter 1977), 65. Marx's fragment in *The Grundrisse,* ed. David McLellan (New York: Harper Torchbook, 1972), p. 45.

[30] Kosik, pp. 73–74: "Memory brings past things out *into the present* and thus transcends the temporary. . . . The historical stages of human development are not emptied forms from which life would have evaporated *because* mankind has reached higher forms of development. Rather, they are continually incorporated into the present through praxis—the creative activity of mankind. . . . Human history is an incessant totalization of the past, in the course of which human praxis incorporates and thus revives moments of the past.

> Unhappy consciousness, tragic consciousness, romantic consciousness, Platonism, Macchiavellism, Hamlet, Faust, Don Quixote, Josef Svejk, and Gregor Samsa are all historically *generated* forms of consciousness or modes of human existence and in their classical form they were created in a *particular,* unique and irreplicable epoch but, once created . . . they themselves form history and acquire a validity independent from the original historical conditions of their genesis.[31]

These forms, that Kosík lists without any distinction, may well be "forms of consciousness" in a phenomenological perspective, but cannot be considered artistic forms, if for no other reason than that they are devoid of materiality. Art and artistic forms cannot be defined or discussed disregarding their material components, the matter of the sign-vehicle.[32] Romantic consciousness, Hamlet, Faust and so on are, rather, literary themes which may survive and reappear in changing forms. Moreover, Kosík's insistence on the permanence of memory and his view of a linear, temporal progression toward "higher forms" of human development are not useful to explain how artistic forms outlive or do not outlive the social conditions in which they were first produced, why some are transformed (e.g., frescoes to contemporary murals, verse epic to prose SF), others have disappeared (silent cinema, the sonnet), and others simply drag on (biographical fiction, much of "children's literature," realist theatre, canvas painting, etc.).

Human reality is in this sense not only the production of the new but also a (critical and dialectical) *reproduction of the old."*

[31] *Ibid,* p. 72.

[32] Eco, *A Theory of Semiotics,* p. 266: "In the aesthetic text the matter of the *sign-vehicle becomes an aspect of the expression-form.*" See the whole chapter 3.7, "The aesthetic text as invention," pp. 261–76. When I say that SF is a form, I see SF as a type of writing with specific mechanisms of construction and inscription, subject to certain constraints of production and reception, textual and material. The term "form," then, is preferable to the term "genre," to which the conventions of criticism assign a restrictive use as internal taxonomic category within established artistic categories like literature or poetry, music or drama. Moreover, the elements and criteria by which genre is determined are often predicated soley on thematic content or chrono-logic patterns in narrative structure as in the work of Northrop Frye, or in Tzvetan Todorov's *The Fantastic: A Structural Approach to a Literary Genre* (Ithaca, N.Y.: Cornell Univ. Press, 1975). The notion of genre, therefore, in assuming these prior categories and usually excluding all material aspects of artistic production and reception, forces the text into the Procustian bed of idealistic or positivistic esthetic categories. Some of the ripples made by Todorov's structuralist proposal of the fantastic as a theoretical genre can be seen in Stanlislaw Lem's critique, "Todorov's Fantastic Theory of Literature," *Science-Fiction Studies,* 1 (1974), 227–37, followed by a set of articles in defense of Todorov and by Lem's response in volume 2 (July 1975), 166–70. Still on the question of generic criticism, see also the debate around David Ketterer's *New Worlds for Old: The Apocalyptic Imagination, Science Fiction, and American Literature* (Garden City, N.Y.: Doubleday, 1974) among Robert H. Canary, S. C. Fredericks, Ursula LeGuin, and David Ketterer in *Science-Fiction Studies,* 2 (July 1975), 130–46.

If the present always reappropriates the past, it does so in complex ways, privileging certain forms, abandoning others, by a process of selection and valorization which cannot be explained simply in terms of human memory. The ways in which, or *how* the past is recognized and appropriated are more directly related to historical conditions than are the "modes of human existence," *what* is being appropriated or sublated from the past. Those ways are the semiotic patterns which shape our processes of cognition and underlie the permanence, the obliteration, or the transformation of forms. Thus the starting point is not the past but the present conditions, as Benjamin saw it: "For it is not a question of presenting literary works in the context of their time, but rather of describing the time which recognizes them, i.e., our own time."[33] History, like SF, has a constantly shifting horizon. And it should not surprise us that SF, in exploring the possibilities and conditions of other modes of social, material and physical existence, is essentially concerned with history. From Asimov's *Foundation* trilogy (1942–51) to Delany's *Dhalgren* (1975) and Russ's *We Who Are About To* (1977), the horizon of history has shifted considerably.[34] Within the historical horizon of World War II, Asimov reappropriated medieval history and the political struggle between the Church and the Empire in a landscape dominated by the technology of metals. The historical horizon of Delany's and Russ's fictions is our own time. That they represent history not as a linear succession hierarchically ordered in discourse, but rather as a discontinuous series of multiply intersecting signs, a discursive construction of reality at once subjective and objective, is only a proof of their historicity. I have tried to show, elsewhere, how Calvino's *Invisible Cities* can be read as a series of semiotic models or cognitive, poetic hypotheses about social economic history and individual reality.[35] Like his earlier works, particularly *Cosmicomics* and *t zero,* this text is written and reads like SF: it generates the reading protocols that I have described as constituting the science fictional mode. While the earlier stories revisited human activity and creativity from prehistory to a future imaginable only in science fictional terms, a history with no privileged moments or individuals, *Invisible Cities,* written in 1972, is concerned with the very possibilities of representation of the historical dialectics of subject and object.

As an artistic form, SF today may have assumed the creative

[33]Walter Benjamin, "Literaturgeschichte und Literaturwissenschaft," *Angelus Novus* (Frankfurt am Main: Suhrkamp, 1966), p. 456, quoted by Peter Uwe Hohendahl, p. 55.

[34]Isaac Asimov, *Foundation* (New York: Avon Books, 1974); Joanna Russ, *We Who Are About To* (New York: Dell, 1977).

[35]See my "Semiotic Models, Invisible Cities," *Yale Italian Studies,* 2, No. 1 (Winter 1978), 13–37.

functions once held by Greek epos and, in the past two centuries, by the historical novel. Like them it is a poetic form of historical representation, a discursive construction of cognitive hypotheses or models of human reality. More than a century has passed since Marx's wonder at the aesthetic pleasure provided by Homer: today such pleasure, as most readers no longer possess the knowledge of classical reading protocols and cultural codes, no longer exists. And the narrative expression-form of the nineteenth-century novel is no longer an appropriate vehicle for the history of "our own time," for the social reality that we must, incessantly, invent.

Generic Protocols: Science Fiction and Mundane

SAMUEL R. DELANY

I

Those of us, such as Professor Suvin and myself, who would maintain science fiction (henceforth SF) is a genre, rather than a local form (and not of a particularly central locality at that), find ourselves confronted with a taxonomic dilemma. Let me elucidate with a number of examples:

One can describe, say, a sonnet with a number of sentences that will be applicable to a vast number of other poems that, in the past, have been clearly labeled sonnets; and these sentences will meet a remarkably rigorous set of criteria involving recognition, replacement, and construction. In short, this description will be of the sort frequently called "a definition"—and, indeed, a definition itself can be described by a number of sentences that meet the same criteria. What cannot be described by such a number of sentences meeting these criteria, however, is poetry itself.

Our second example: one can describe, say, a locked-room mystery in such a way that the description will fit many formerly so-labeled fictional works; and again, we can put such a description in such a form that is recognizably a definition. What we can not describe in such a manner is prose fiction.

And our third example, purposely a non-verbal one: we can

formulate a definitional description of a "still-life" that applies to many paintings, drawings, photographs, and other examples of the visual arts. But it is "the visual arts" that cannot be defined.

The sonnet, the locked-room mystery, and the still-life are all local (or, more accurately, locable) forms.

Poetry, prose fiction, and the visual arts are all, of course, genres.

The taxonomical dilemma is simply this: as soon as we would take a group of texts, a recognizable type of writing, say, till now considered simply "a local form," and argue that it is really a genre, we are in the odd position of having to describe in a formal way why these texts can no longer be formally described, and then go on to describe them informally in such a way that their processes, operations, functions, and general workings are still clearly recognizable.

Inductively, we all have a pretty good sense of the "indefinability" of science fiction. In both the fanzines and the scholarly journals, we have watched definitions posed and definitions refuted. All have seemed too broad or too narrow. But induction, of course, runs counter to the formality we are seeking. Who is to say that a counter example will not surface tomorrow. The only real way to get beyond the problem is to examine just why genres resist such descriptions. Exactly what sort of objects are they?

The reason genres are not rigorously definable is because they are not simply constituted of a set of texts (and for those who wish to keep the non-verbal aspect of our agreement open, we will describe texts here with the informal description given by Gayatri Chakravorty Spivak: "A text is anything constituted of a play of presences and absences,"[1] the usual substantive for presences and absences of course being *meanings*).

A genre, to the extent that it can be said to exist as a genre, is constituted of a way of reading (what we may henceforth call a protocol of reading), a structuration of response potential. A set of texts, over a period of time, to the extent they are different from other texts, produces a way of responding to these texts that is different from the way we respond to texts we might say belong to other genres. Still other texts, intended to be part of the genre, will be written in such a way as that particular protocol allows them to "make sense" or make a richer sense than some other generic protocol would. And still other texts will require that this protocol be slightly shifted in order to produce richer readings. (Or possible weaker ones.) Thus the genre changes over the years.

But it is precisely this aspect of genre—generic existence as a protocol of reading—that allows us, say, to explore the "poetic" facet of a work of prose fiction, or the "dramatic" aspect (for certainly drama is one of the

[1] As stated in a talk given at the University of Wisconsin-Milwaukee on 10 November 1977.

irrefutable genres) of a poem. When we take a piece of prose fiction and apply the reading protocol we would normally apply to a poem, we pay more attention to the sound of the words; and we begin to interpret the text by a whole set of informal codes referring to various emotional affects usually associated with poetry. That poetry *is* a genre—a protocol of reading—is what allows us to perceive the poetic aspect of texts not properly poems. In the same way we might pay attention to the fictive aspects of a text that is, properly, a poem: We pay particular attention to the verbal signs indicating persons ("characters," prose fiction calls them), the accuracy of certain social or economic descriptions, etc. In short, we might read a poem by the protocol which, as it has grown up in a response to many texts, has become the genre we call fiction.

Similarly, we can look at the dramatic aspect of either a poem or a piece of prose fiction—which may mean anything from paying more attention to the use of quoted or reported speech, or a certain dialectical organization of argument, to a certain particular set of emotional affects having to do with conflict or excitement associated with the dionysiac aspect of theatrical climaxes.

The fact that each genre is a way—a protocol—of reading (formed over a historical period, certainly, by many texts), is what allows that protocol to be applied to any number of other texts as well as what prevents a genre from being defined simply by a description of the texts themselves.

In the past I have written at varying lengths about the particular reading protocol that is science fiction in its generic aspect—the reading protocol that allows us to interpret particular sentences in such a way that, at the level of the signified, they are clearly part of SF. I will give the briefest review here:

Then her world exploded.

Should such a string of words appear in a mundane fiction story, we would more than likely read it as an emotionally muzzy metaphor about the specifically emotional aspect of some incident in a female character's life. In SF, we must retain the margin to read this sentence in such a way that a planet, belonging to some woman, blew up.

He turned on his left side.

Mundane fiction would more or less constrain us to read such a string of words as referring to a man's insomniac tossings. In SF we must retain the margin for reading such a sentence as: some man reached down and threw a switch activating the circuitry of his sinistral flank.

And there are many other sentences in SF with a perfectly literal and lucid meaning (e.g., "The door dilated," from Heinlein's *Beyond This*

Horizon) that, read as mundane fiction, would be meaningless at worst or catachresis at best.

To learn to read these and like sentences as meaningful in the context of an SF story is to learn a significant portion of the reading protocol that is the SF genre. Consider: there is no sentence I can think of that theoretically could appear in a text of mundane fiction that could not also be worked into some text of science fiction—whereas there are many, many sentences in science fiction that it would be hard or impossible to work into a text of mundane fiction. Science fiction discourse gives many sentences, that in mundane fiction would be meaningless or at any rate very muzzily metaphorical, clear and literal meanings. Just at the level of lucid and literal sentence, then, which is the larger way of response, the wider range of understanding? Which offers the greatest range of readings for possible sentences?

More recently I have been exploring the way we actually organize the information from science fiction texts—exploring the organization principles of science fiction discourse. Because in the discourse of mundane fiction the world is given, we use each sentence in a mundane fiction text as part of a sort of hunt-and-peck game: all right, what part of the real world must I summon up in imagination to pay attention to (and, equally, what other parts—especially as sentences build up—had I best not pay attention to at all) if I want this story to hang together? In science fiction, the world of the story is not given, but rather a construct that changes from story to story. To read a science fiction text, we have to indulge a much more fluid and speculative kind of survey. With each sentence we have to ask what in the world of the tale would have to be different from our world in order for such a sentence to be uttered—and thus, as the sentences build up, we build up a world in specific dialogue with our present conception of the real.

Again, to take on a string of metaphors that, alone, might lend itself to either discourse, here is a sentence from Pohl and Kornbluth's *The Space Merchants:*

> I rubbed depilatory soap over my face and rinsed it with the trickle from the fresh-water tap.[2]

In a mundane fiction tale (i.e., if we were to read such a sentence by the protocol of mundane fiction), since the world of the mundane tale is a given—some part, presumably, of the real world—we would read the adjective "fresh" as either a redundancy (since, in that world, the vast majority of faucets *are* fresh water faucets) and therefore a writerly failing,

[2]Frederick Pohl and C. M. Kornbluth, *The Space Merchants* (New York: Ballantine Books, 1953), p. 1. This particular example was first pointed out to me by Tom Moylan, University of Wisconsin-Waukesha.

or possibly as a comment on the consciousness of the point-of-view character (perhaps for some subjective reason he is abnormally aware of the water's freshness); by extension, we would read the "trickle" as there either to support, or to contrast with, this particular state of mind through some metonymic connection between freshness and abundance. In an SF story—indeed, in *The Space Merchants*—while all these suggestions, muted, are there, the sentence is doing a radically different job at the level of the signified: in SF, the world of the story is not given; rather it is constructed from such sentences, and in the world of *The Space Merchants,* because of the over-population houses have both fresh and salt water faucets. The trickle indicates that, in this world, the fresh-water supply is low, even in a luxury apartment complex like the character Mitch's. In mundane fiction, any expression at odds with a given normal world picture becomes information about the internal state of the character observing the world. In SF, we must retain the margin for reading every expression away from a given norm as informing us not about the fictive character so much as it informs us about organization of the fictive world.

There are many people who simply "cannot read" SF. (And I distinguish these very clearly from the people who simply *will* not read it out of snobbishness: though no doubt the two categories overlap.) A number of times now I have had a chance to read some science fiction texts with such readers sentence by sentence, both students and faculty, at the two universities where I have taught courses in SF, and it seems that when there is a difficulty in understanding, it falls precisely at this level: the little hints, suggestions, and throw-aways by which the most skilled science fiction writers weave the tapestry of their world into coherence, just do not start the proper responses in these readers. For SF, as a genre, *is* a language, a set of conventions, a reading protocol—and it must be learned by exposure and application.

At any rate, these distinctions in reading protocol that we have just talked about are the stuff of generic distinctions. Indeed, one would be hard-pressed to discover such sharply distinctive examples between, say, poetry and mundane fiction, or prose fiction and drama. But what we have been outlining here is the way which the little sign "SF" in the upper right-hand corner of a paperback original novel is not a simple metonym for a set of fictive conventions that knot together into a submerged net, totally describable, even if it contain infinite gradations, by a circumscription of referential topics, or if it is infinitely extendable, by an algorithm into which one need only feed the proper social variables (new advancements in science, say), any one or any several of whose hidden nodes may surface in any given SF text, the bulk structure of this net still totalizably cognizable; it is rather one of many possible switches that bring into play nothing less than a protocol of reading by which an infinite number of sentences, phrases, and even coined words may be meaningfully interpreted, themselves only an infinite subset, part of and equal to the infinite set of sentences, words, and phrases, that is the language of which they are a part.

II

Reading seems to be a mental field of response that a history of exposure to various written statements contours toward a structuration which *is* understanding. The genres of reading seem to be the response locus of various historically contoured groups of texts which we are always free to apply outside of those histories—the reading protocols with which we try to make history our own.

There is a view which reads "reading" as a perfectly transparent gaze that simply responds to what is placed before it with a mythical homuncular institution called "understanding." And what is placed before it is, of course, writing. This view that identifies writing with understanding sees reading as a transparent medium connecting the two in a situation of perfect closure. It sees all problems of comprehension as lexically embedded: know the meaning of the word and you know the meaning of the sentence; know the meaning of the sentence and you know the meaning of the text. It takes no cognizance of the fact that at every level—word, phrase, sentence, sentence cluster, paragraph, topic, text—economic distortions are at work, sensitive to every other level at once: any sum of these may be a genre; these summations are not unique; and any alternate summation may be an alternate genre.

I have labeled this particular view (or lexical extension and readerly transparence) "the metafictive fallacy"—after one of the most prevalent emblems by which it betrays itself. That emblem is the notion that "all writing is a critique of former writing protocols." The historical assumption behind this expression of the fall of language is fairly simple: somewhere at a more or less locable point, writing—fiction—managed to inscribe directly, with a minimum of mediation, the social reality of cultural life. (When this moment is identified with the "classical," then the absence of mediation is usually seen to lie not just between society and writing but also between every other social institution as well, i.e., it constitutes the presence, or the integration, that classical writing "reveals" classical society to have possessed.) Over an historical period, however, writing begins to reflect not life, but rather the vaguest shadow of the social texture (precisely to the extent that the society itself falls away from its moment of classical integration), heavily and hopelessly mediated by the structures of the classical text. Thus the writing is contaminated in three ways: first, in the classical text, all is immediately expressive of the classical society's unity; but since society is no longer unified, all these classical holdovers now move away from the immediate (in terms of the current society) to become mediation (i.e., distortion) itself. Second, because our society is no longer unified, we no longer know how to use the classical holdovers immediately (even though, somehow, through that transparent reading their expression of a vanished unity is still perfectly apparent). Third, though the use of classical holdovers in modern texts lends a patina of classical luminosity, we cannot read them properly (reading is transparent and gazes directly onto life) because our lives are not classical so there is no classical context to charge them with their true meaning. (Does the logic of these contaminations contradict? But we are outlining the structure of a fallacy.)

The logical outcome of this set of presuppositions is that modern writing only becomes responsible for and to itself when it becomes aware of this situation and conscientiously takes over for its central field of play the conventions of past writing and abnegates any attempt to respond directly to current "life/society." For this is impossible due to those past and opaque conventions which are at once perfectly expressive of classical unity and at the same time dense to any real understanding since we no longer live in a classical world (and therefore they no longer have a classical context)—those conventions completely mediating our view of the present reality so that we can no longer respond to it immediately. Thus, the only true subject for modern fiction is past fiction—i.e., what critics, since Scholes coined the term in a parodic gesture towards Jameson's "metacriticism," have been calling "metafiction." The final conclusion to all this is simply that, however clever, entertaining, sympathetic, or profound it appears, this modern fiction is always a fallen fiction and is trivial before the previous fiction whose classical unity is always at once visible and invisible.

To reiterate: This position seems to hinge on what I have termed the metafictive fallacy—though I call it after the fruit it bears, not after the organization of its roots.

To return to those roots: reading is transparent, which allows those moments where the inescapable opacity of reading becomes momentarily manifest to bear a logical relation to that supposed transparency, no matter how circular or contradictory.

What I would pose in place of this fallacy is simply that reading *is* that which is mediation itself; and is therefore always mediate. What I would propose as a closer approximation of the real is that reading is not transparent before the text, but rather is our only knowledge of the text: and corresponding to the point in the fallacious system where we find the notion, "All writing is a critique of former writing protocols," I would substitute, "All writing is a critique of the current reading protocol." The reading protocol must be conceived of, in both its density and its malleability, as conditioned by far more than simple patterns of previously written language. (Without denying the part previously written texts have in contouring this protocol, we must note that current speech and whatever social reality we have experienced, however we managed to experience it, mediated by whatever, to whatever extent, must play their part too.)

Narrative seems to be able to critique the reading protocol in at least three basic ways: the narrator can present to the reading eye topics never (or rarely) referred to before (a newly mediated reference). The narrator can write of old topics but positioned in such a way that the economic distortions of the medium force new valuistic interpretations (old topics re-mediated). Or the narrator can write of new topics re-mediated—this last being by far the most difficult because it demands the clearest vision of the mediation process that already exists, and should probably only be taken up by writers with not only an experimental bent but some real analytical genius.

A genre, then, is a communal entity with the ontological status of Saussure's *langue,* Durkheim's society, or Freud's unconscious. It is far closer to a statistical construct than it is, say, to a drawing from life. It is only normal in the mathematical sense, and for that reason, judgementally useless—and for *that* reason is almost always employed with ideological abuse.

If our supposition that reading is not transparent and unyielding but rather opaque and responsive is true, then we may go on to suppose that in any genre we can assume the following: a genre, a reading protocol, is formed by (among other things) a set of texts. New texts intrude on this protocol (some labeled SF, some labeled "Scientific American article," some labeled "literature," some labeled "criticism," some labeled "real life," and some labeled "pure fantasy," among many, many others, labeled and unlabeled) to change it.

III

At Oxford in 1892 the French poet Mallarmé delivered a lecture that began with the now famous line, *"On a touché au vers"*—someone has been tampering with poetry. Today, eighty years later, I had thought of beginning, "Someone has been tampering with science fiction." But if I did, I would have to make some distinctions between 1892 and 1978 right off. For one thing, in 1892 the person who was doing (by far!) the most tampering was Mallarmé himself—along with a few poets who were comparatively closely associated with him (they came for coffee every Tuesday evening).

The tampering I'm talking of is not coming from within science fiction. When I read writers who are just my juniors, in length of time published if not in years (John Varley, James Tiptree, Jr., Michael Bishop, Vonda McIntyre, Jean Mark Gawron, Suzy McGee Charnas, or Joseph Haldeman, to name the most random few), though of course I see local disagreements and a whole variety of different approaches to the world between them and me, between each of them and each other, I don't sense any violent rupture between these newer writers and those writers who are my immediate contemporaries (Disch, Le Guin, Niven, Russ, Zelazny, to name another random few).

The tampering I'm talking about does produce a sense of rupture. Though there is much disagreement among writers of all generations whether this rupture is a good or a bad thing, we all sense it. It is the tampering that comes from academia, from critics who have become "interested" in science fiction. We have to locate why this tampering is experienced as rupture and as encounter—and I don't mean simple prejudicial xenophobia. Having had a chance to teach science fiction at two universities in the last few years, as well as a chance to write my share of academic criticism and survey the present academic response to science fiction,[3] I'm in a particularly good position to experience the rupture aspect—and yes, it *is* an experience.

In 1975, when I was organizing a scholarly symposium on science fiction at SUNY-Buffalo, I was extremely excited to have in attendance an exemplary Joyce scholar and literary theoretician who was about to publish a book on science fiction with a pollysyllabic title from a highly respected university press. The day the symposium began, advance copies of the book arrived! I made a breakfast appointment with this very affable gentleman to discuss his book with him the next day—and stayed up till four o'clock in the morning reading over the book twice and filling the margins with notes and comments. Over scrambled eggs and toast, I gave him my notes: they ranged from proofreading errors, to correcting dates, to respectful deferments on matters of opinion. But at one point, I referred to something he had to say about the use of matter transmission in science fiction, using Niven's *Ringworld* for his example. His idea had to do with "matter transmission" as a metaphor for a vision of "telekinesis," and what he felt telekinesis meant to people. "I'm just curious," I said, "Why, if you wanted to make a point about telekinesis, you didn't refer, say, to Alfred Bester's *The Stars My Destination,* where the idea is dealt with directly, and in very much the manner you outline. Do you think,

[3]You can find the results of this tampering in two new critical books by me: *The Jewel-Hinged Jaw* (New York: Berkley-Windhover Books, 1978), a collection of essays in paperback, and *The American Shore: meditations on a tale of science fiction by Thomas M. Disch—Angouleme* (Elizabethtown, N.Y.: Dragon Press, 1978).

perhaps, the book has received too much attention? Or perhaps it's not as good as people are always going as if it were?"

And this gentleman, who had been writing so eloquently about Le Guin's themes and Sturgeon's prose, looked at me with perfect ingeniousness and asked: "Bester? *The Stars My Destination?* Is this a book or an author I should have heard of before?"

This is totally disorienting; it throws the whole discussion into a level of surrealism that is mind-boggling. Someone who writes a book on a subject, about whom you can say "They don't know the field," is usually someone who gets the dates wrong, forgets small facts, comes to wrong-headed opinions. Perhaps there are a number of important works that they haven't read recently enough or closely enough and therefore are relying too heavily on what another writer had to say about them. But can you imagine asking someone who has just written a book on twentieth century poetry why they didn't mention T. S. Eliot or *The Waste Land,* only to get the perfectly serious answer: "T. S. Eliot? *The Waste Land?* Is this a poem or a poet I should have heard of before?"

This is rupture.

The particular critic, I'm happy to report, over the next two years did a lot of homework and wrote a much better book with a much less polysyllabic title—published by a different university press.

But the experience of rupture remains.[4]

Then there was the academic critic who, from his chapter on science fiction in his book on fantasy, had apparently discovered Michael Moorcock's delightful *War Lord of the Air,* and decided that Mr. Moorcock had, out of sheer original genius, invented an entirely new sub-genre of science fiction, which he dubbed, "the historical alternative story." He went on to say that though he suspected there would be a lot of argument among regular SF fans about whether Moorcock's brand new SF twist would be accepted or not, he felt this new form really should be accepted into the overall genre of science fiction—just as if Dick's *The Man in the High Castle* had never been written, nor won its much deserved Hugo award for best novel of its year, not to mention his complete ignorance of all the other parallel world stories (historical alternative indeed!) from Ward Moore's *Bring the Jubilee* to Hilary Bayley's "The Fall of Frenchie Steiner" and Joanna Russ's *The Female Man*.

[4]Of course, such rupture as we experience it at the hands of academics is not new. We've experienced it before at the hands of editors and publishers who really *do* have their hands on our economic jugular veins. And we've survived it—survived it very well. In 1951 (the year of the establishment of the MLA Continuing Seminar on Science Fiction), there were some fifteen texts published that could reasonably be called science fiction novels—including the serials in magazines and the first volume of the "Foundation" series, which was really a compilation of stories written since 1942. Last year, over fourteen percent of *all* original fiction published in the United States was science fiction.

The rupture I experience is a rupture with my own history, with my own knowledge of the historical reality of science fiction writing, a rupture with the genre itself. The working assumption of most academic critics, an assumption which certainly distorts what they have to say of specific texts, is that somehow the history of science fiction began precisely at the moment they began to read it—or, as frequently, in the nebulous yesterday of sixteenth and seventeenth century utopias. For both accomplish the same thing: they obviate the real lives, the real development, and finally the real production, of real science fiction writers, a goodly number of whom are still alive if not kicking. To assume that something—like science fiction—has no *significant* history in the past is to assume that its history-to-come will be no different from the last phenomenon whose history you've been studying. I don't mean in terms of dates and occurrences, but in terms of values, processes, ways of understanding it and responding to it. To say that a phenomenon *does* have a significant history is to say that its history is different from the history of something else—that's what makes it significant. To say that a phenomenon has *no* significant history at all is a very sneaky way of allowing yourself to treat it as *if* its history were exactly the same as that of some other phenomenon you are already acquainted with. And the historical phenomenon most literary critics are most acquainted with is, of course, literature.

After we have passed the sense of rupture, here is where we locate the sense of encounter. And it's the growing number of feet of shelf space in bookstores, the growing number of readers who turn to science fiction, the growing number of reading hours that readers are devoting to science fiction, and the growing number of courses given on science fiction in the country's high schools and universities (over five hundred at last count) that give this encounter all the attributes and urgency of a battle.

Myself, I find science fiction's literalization of the language, its wealth of clear and lucid sentences, simply and sensually pleasurable. I find the dialogue it sets up with the real world (a dialogue that mundane fiction simply cannot indulge) both pleasurable and useful—if only because it keeps the possibility of dialogue alive. But if we really want to explore the encounter between values that, finally, *is* the encounter between literature and science fiction, we have to go into the values of literature as well.

The French scholar Michel Foucault is one of the most radical and fascinating thinkers to tackle this problem. In an essay called "What Is an Author," he notes that many of the values of literary discourse are tied up in the very concept of the "author" of a work. The author, or, as he sometimes calls it, "author-function," becomes the focus for some of literature's most central values. In this essay he writes:

> The manner in which literary criticism once defined the author—or rather constructed the author, beginning with existing texts and discourses—is directly derived from the manner in which Christian tradition authenticated (or rejected) the texts at its disposal. In order to "rediscover" an author in a work, modern literary criticism uses ways similar to those that Christian religious commentary employed when trying to prove the value of a text by its author's saintliness. In *De Viribus illustribus*, Saint Jerome . . . proposes four criteria: 1) if among several books attributed to an author one is inferior to the others, it must be withdrawn from the list of the author's works (the author is therefore defined as a unified level of value); 2) the same should be done if certain texts contradict the doctrine expounded in the author's other works (the author is then defined as a field of conceptual or theoretical unity); 3) one must also exclude works that are written in a different style, containing words and expressions not ordinarily found in the writer's production (the author is here conceived as a stylistic unity); 4) finally, one must consider as interpolated those texts which quote people or mention events subsequent to the author's death (the author is here seen as a historical unity and the crossroads of a limited number of events).
>
> Modern literary criticism, even when—as is now customary—it does not concern itself with authentification, still defines the author no differently . . . the principle of a certain unity of writing—all differences having to be resolved, at least in part, by the principles of evolution, maturation, or influence.[5]

Clustered around the literary concept of "author," then, we find this quartet of literary values: unity of value, theoretical unity, stylistic unity, historical unity. It is a little sobering to consider that a discipline like literary criticism grew so directly out of a dogmatic religious enterprise. But these values are certainly among its controlling parameters. One of the last major battles in the history of the English novel was the furor over whether or not D. H. Lawrence was to be accepted as a Great Author, or consigned to the category of interesting crackpot. The critic F. R. Leavis, in his *D. H. Lawrence: Novelist*, that pretty much settled the question, goes out to prove Lawrence's greatness, right in chapter one, by showing the "unity" of Lawrence's works.[6]

And I have seen at least one master's thesis written about my own science fiction that set out to prove me an author worthy of serious consideration—by demonstrating the unity in none other than the works of Samuel R. Delany!

[5]I have used, and very modestly revised, Josué V. Harari's translation which is included in *Textual Strategies: Perspectives in Post-Structuralist Criticism*, ed. by Josué V. Harari (Ithaca, N.Y.: Cornell Univ. Press, 1979). Another version of the Foucault lecture may be found in Michel Foucault, *Language, Counter-Memory, Practice*, ed. Donald F. Bouchard (Ithaca, N.Y.: Cornell Univ. Press, 1977), pp. 113–38.

[6]F[rank] R[aymond] Leavis, *D. H. Lawrence: Novelist* (Cambridge, England: Minority Press, 1930).

At this point we have to ask: Are these unities part of science fiction discourse? Should they be applied to science fiction?

I've already talked about the way, sentence by sentence, science fiction can differ from mundane fiction. I've talked as well about the way science fiction organizes information—not only into a story but also into a world—that is different from the way mundane fiction organizes its information. I also feel that if we look for this quartet of literary unities—valuative, theoretical, stylistic, and historical—in science fiction discourse, whether clustered around the author or not, we will find diametrically opposite values.

Working backwards through them: One must consider as inauthentic "those texts that quote people or mention event's subsequent to the author's death." Well, that certainly lets science fiction out of the historical-unity game! Science fiction's very commitment to its future vision means that the SF writer is always quoting people and mentioning events subsequent to the writer's death! So this basic image of historical unity is denied at the outset. But it's not the image we are concerned with so much as the value as an operative function: and the historical value science fiction seems to operate by, more than any other, is one of historical plurality—a value diametrically opposite to the unitary value of literature. This is reflected not only in the diverging historical views within the production of a single writer (nothing stops me from writing three science fiction stories, all set in New York City in Twenty One Hundred, one in an over-populated world, one in a depopulated world, and one in a world whose population has managed to stabilize at, say, two and a half billion: they would simply involve three different extrapolations), but also the parallel universe tales that so astonished the academic about whom I wrote earlier.

This is possibly the place to point out that the author, or author-function, simply plays a very different role in science fiction discourse from the one it plays in the discourse of literature. I doubt I have ever called myself a "science fiction author"; the term would simply feel too uncomfortable in my mouth. When someone asks me my profession, I say I'm a science fiction writer. By and large readers tend to be much more concerned with stories than with writers. But this leads us to the next value.

The value of unity of style: science fiction's origins in the pulps and persistence as a generally popular literature simply mitigates against the sort of stylistic unity that literary value demands, both in the productions of single writers and, certainly, in the production of the whole field. Writers are always adopting different styles for different stories, and evolution, maturation, or even influence are just not the operative factor—the stories, or even the particular level of the readers, demand them. For a good long while now science fiction has been responding to readers on all levels: someone who loves the simplistic thrust of a Perry Rodan book is probably not going to love the techno-social recomplications of a John Varley or the logico-linguistic invention of a Jean Mark Gawron—though I know of at least one mathematics professor who reads all three avidly. The point, however, is that all three are science fiction. But because of the range of markets, the range of readers, there is simply very little chance of stylistic unity as we find it in the literary concept of author-function. Again, if anything, there seems to be a highly valued ideal of stylistic plurality—especially since the science fiction of the sixties.

Well what about theoretical unity? The other side of science fiction's commitment to historical plurality is an equal commitment to theoretical plurality. What has most confounded the folks searching for definitions of science fiction in terms of scientific subject matter is the number of science fiction stories that clearly contradict known science—take all the stories with faster-than-light travel, for example. Then, of course, there are all the undeniably science fiction stories about magic (e.g., Gogswell's "Wall Around the World," Blish's *Black Easter*). To say, well, in these tales magic is treated in a "scientific way" only blurs the question: currently the existence of magic runs counter to scientific theory, and that's all there is to it. Then there are all the stories about ESP, which, if not exactly contradicted by prevailing theory, is certainly rendered highly dubious by it. Mumbling about "exceptions that prove the rule," whatever that means, simply doesn't cover the case. The concept of theoretical plurality, as an operative value, does. For there to be such a value, the genre, across its range, *must* deal with conflicting theories. This value does not necessarily fix itself to the "author" function in science fiction: not every writer feels the necessity to choose an opposing theoretical construct from tale to tale—though many have. I would hazard, however, that every science fiction writer, precisely to the point that her/his own work is theoretically consistent with itself, is very clearly aware of one or more science fiction writers with whom that theory conflicts, whether the theory be political, sociological, or scientific.

Finally there is unity of value itself. As history and theory, whether unitary or plural, form two sides of a single coin, so style and value, whether unitary or plural, form two sides of another (and here, of course, style means a little more than merely use of words; there are styles of thinking, styles of perception). The same factors that assure SF will not exhibit any unity of style in the literary sense, but rather a plurality of styles, both within the production of single writers as well as throughout the generic range, also means that science fiction actively strives for a plurality of value (i.e., worth). Once a text is adjudged "literature," we can say it partakes of a certain (admittedly vague and almost impossible to define) value, a value that, however vague, consists of a juxtaposition of theoretical, stylistic, and historical elements. This value—the text's literary value—militates for the text's preservation, its study, its reproduction. But once more, this is *not* the case with science fiction. Having adjudged a text science fiction, we simply have made no unitary statement, however vague or at whatever level of suggestion or implication, about its value. Again, I suspect this is because inherent in the discourse of science fiction is the concept of value plurality.

It may be well to point out here exactly what we have done—so that no one is tempted to over-value *our* exploration. We have simply taken the list of values Foucault has recovered under the literary concept of "author" and let them guide us through the range of science fiction—whereupon we found some values that pretty much oppose the literary ones. We have not necessarily discovered the *most* important values of science fiction. *They* may be completely otherwhere. The ones we've found only take their particular highlighting when held up against the literary.

So: do I feel that science fiction will, or should, be taken over by literature in the current encounter? I sincerely hope it is not. And the only way I feel it can be taken over is for very bad academic criticism—the kind that strips fiction of its history, that ignores it as a discourse, as a particular way of reading and responding to texts, and that obscures its values of historical, theoretical, stylistic, and valuative plurality—to swamp what I feel is a responsible academic approach, of which I offer my own argument up till now as a modest example.

This brings us to what may well be the most important battlefield in the encounter: around every text there is a space for interpretation. There is no way to abolish the interpretive space around the text: it comes into existence as soon as we recognize the words' meanings.

IV

Locating the play in the interpretive space (rather than a unitary or hierarchical explanation) can be done in a number of ways.

In the song from Shakespeare's play *Cymbeline*, we find the lines:

Golden lads and girls all must
As chimney sweepers come to dust.

It seems a pretty clear statement about the eventual death of even the young and beautiful as well as the dirty and grubby. Some time in the thirties, a scholar traveling in Warwickshire, the county of Shakespeare's birth, discovered that the local slang for the flowers we call dandelions was "goldenlads"; and when the yellow fuzz was blown off the dandelions' heads, they were called "chimney-sweepers." And, apparently these slang terms are several hundred years old. Read the two lines again. They haven't *lost* any of their meaning. But a whole range of play has been introduced with the recovery of the local Warwickshire dialect. If one wants to be "literary" about things, one can hierarchize *all* the meanings into a logical unitary order to turn them into a single, coherent essay—indeed, as we have seen before, we can turn them into several different coherent essays and then (if you want) begin all over again, hierarchizing *them*. I would hazard that Shakespeare's delight in the line, as well as the delight of his audience, was in the simple play of plural meanings that we now have, knowing both literal and dialect interpretation of the terms.

What does this little diversion have to do with science fiction? Well, when Roger Zelazny, in *This Immortal,* writes of a biologist breeding poisonous fleas, called slishi, to kill off an invasion of spiderbats on the Monterey coast, "When the spiderbats return to Capistrano, the slishi will be waiting," he is basically initiatig the same sort of play as Shakespeare. But to perceive the play, one must know that there was an extremely sentimental old lyric, "When the Swallows Come Back to Capistrano." Zelazny's line puts that sentiment in play with the grim literalness, and the result is amusing, entertaining, and though highly suggestive, does not really lend itself to a unitary, single interpretation. (If we may add play to play: it seems that the play's the thing. . . !)

Here's another way in which historical awareness can indicate the play in both a "literary" writer like Shakespeare and a science fiction writer like . . . Asimov! We know, for example, from historical research, that Shakespeare's plays were performed with elaborate costumes—and *no* scenery at all (if you don't believe me, check Asimov's two-volume *Asimov's Guide to Shakespeare*).[7] This is why the characters spend so much time describing where they are, in ways that, if a cowboy in your latest movie-western did it ("Well, here I am in this dark wood full of elms and sycamore, as the light dims and the pinecones cast long shadows over the dead leaves around my boots."), would make the audience howl. To know this today allows us to read these parts of the dramas in a context that lets them do their jobs again; and it lets us respond to the many subtle ways in which descriptions of locations are worked in, rather like the little throw away bits that give you the world of a science fiction story, even in the midst of character interchanges. They no longer seem grey, awkward, and superfluous.

[7]Isaac Asimov's, *Asimov's Guide to Shakespeare* (Garden City, N.Y.: Doubleday, 1970). The good Dr. A has also written a two-volume guide to the bible: *Asimov's Guide to the Bible* (New York: Avon, 1971); as well, he has written an annotated version of Lord Byron's comic epic, *Don Juan: Asimov's Annotated Don Juan* (New York: Doubleday, 1972).

Notice that all this information, when written into the interpretive space around the text (whether it is Shakespeare's text, Asimov's, or Zelazny's), results in the text's becoming vivider. *More* things can go on in the text. The information is not used to constrain the text to a single, or unified meaning. Rather, in each case, it releases meanings which then come into the play of meanings that is the text. (Think of "play" not so much in terms of children's fun or adult competition, but in terms of a gear or a steering wheel that has play in its movement—though certainly all those other meanings represent points about which the play—in the word play—moves, as does the whole idea of theatrical play as well.) Notice this is *not* the same thing as saying, "The text can mean anything you want. . ." with its implication, "Choose whichever *one* you prefer," which gets us back to the unitary.

This seems to me to be, with both literary texts *and* science fiction texts, the proper use of the interpretive space that lies about them both. Many science fiction readers, however, confuse the existence of that interpretive space about the text with the values those interpretations most often written into that space have, most often, supported—those literary values that are unitary and authoritarian. The response of these readers—frequently our older readers—no doubt impelled by the best of intentions (their wheedling suspicion of the inappropriateness of unitary values to a genre that is so clearly a pluralistic enterprise), is simply to deny the interpretive space about the science fiction text. The usual way of accomplishing this is for these readers to assume a conscientiously philistine approach—which is what they intuitively feel is opposed to a "literary" approach:

"First of all," they say, "science fiction is merely entertainment."

This, of course, is not just a single pronouncement that ends there. It is part of a whole, philistine reader view, and is associated with a whole galaxy of pronoucements. Anyone who has been around science fiction for any length of time will recognize that they all go together.

"I like a science fiction story that's told in good, simple language with none of your fancy writing or experimentation, with a nice, clear beginning, middle, and end."

But haven't we encountered, on the level of values, something very like this? Of course. It's nothing but an appeal for a unity of style.

"I like a science fiction story that sticks to good, hard science that we can all understand if we just know our general physics and chemistry."

But on the value level, we should recognize this one too: it's the call for theoretical unity, loud and clear.

"I guess I just wish they would write science fiction stories the way they did back in the 60s/50s/40s. . ." (You can choose your decade: there are adherents to all of them today.) You guessed it: it's the cry for historical unity.

Paradoxically, it is just this most philistine of reader reactions that most strongly encourages the overthrow of science fiction by literature—because it writes itself in the space of interpretation (and the philistine interpretation of science fiction is no less an interpretation than is the notion of science-fiction-with-no-significant-history still a historical notion) that, through a process finally not too far from bad academic criticism, has very little awareness of the structure of science fiction discourse, either as a historically sensitive process or as a present reality, in which each present writer is inserting her or his play into the plurality of values—valuative, theoretical, stylistic, and historical—around which our discourse moves. For we are not talking about complexity, or even quality, of interpretation, but rather about the values that a whole range of interpretation, good and bad, simple and complex, reinforce. And the philistine view is right there, with all its authoritarian vigor, at the center of the literary enterprise—even when it may well be the play of pluralities that the person expressing that view is actually responding to in any given science fiction text that delights.

V

What this essay has been on the verge of proposing, as some of you by now no doubt have suspected, is nothing less, in this encounter, than the take-over of literature by science fiction. This has been suggested (with varying degrees of play) by various writers at various times in the past. But it is just what gives the phenomenon its aspect of encounter that also, today, in a very real way makes that a possible outcome. Again, I do not mean an economically encouraged encounter between texts: texts labeled science fiction driving texts labeled literature off the shelves. Even the rise from practically zero percent to fourteen percent of fiction production in twenty-five years (nor the rise from zero to about five hundred classes) does not seriously threaten the production of texts of the sort we call, say, mundane fiction or poetry. There are too many other economic pressures, academic and journalistic, that would certainly bring the process to a grinding halt at fifty-fifty, if not well before. I am still talking about the encounter between discourses, between responses, ways of reading texts, ways of using the interpretive space around them.

There are many people who read only literature.

There are also many people who read only science fiction.

But there are also people who have moved from one to the other. The label "silly kids stuff," so long applied to science fiction, was there to suggest that the natural and healthy movement over the period of maturation was from science fiction *to* literature, with its concomitant suggestion that any movement in the other direction implies mental softening. But of course there *are* many people recently who have moved in the other direction—another expression of the encounter.

I talked to one such man, not long ago. A historian, specializing in the beginnings of the nineteenth century, he had been a great reader of literature, but had found, over a period of five or six years, that he was reading more and more science fiction until, for the last two years, other than his journals and non-fiction, he had read nothing else. "I was really afraid to go back and read a 'serious' novel," he told me. "I didn't know what would happen. Finally, in fear and trembling, I picked up Jane Austen's *Pride and Prejudice,* always one of my favorites, just to see what happened when I did. . . Do you know something? I thoroughly enjoyed it, more than I ever had before. But I realized something. Before, I used to read novels to tell me how the world really was at the time they were written. This time I read the book asking myself what kind of world would have had to exist for Jane Austen's story to have taken place—which, incidentally, is completely different from the world as it actually was back then. I know. It's my period."

As far as I can tell, this man has started to read Jane Austen as if her novels were science fiction. There has been an encounter.

Selected Bibliography

The following bibliographical selections are offered for further reading in the theoretical and critical directions outlined in the three sections. They are not meant to provide a bibliographical introduction to the respective fields of criticism nor do they give a panorama of current research.

I. *TECHNOLOGY AND AMERICAN CULTURE*

Bagdikian, Ben H. *The Information Machines.* New York: Harper & Row, 1971.

Bell, Daniel. *Toward the Year 2000.* Boston: Beacon Press, 1969.

Boorstin, Daniel G. *The Americans: The Democratic Experience.* New York: Vintage, 1974.

Ehrlich, Paul R. *The End of Affluence.* New York: Random House, 1974.

Ferkiss, Victor C. *Technological Man: The Myth and the Reality.* New York: Signet, 1969.

———. *Future of Technological Civilization.* New York: Braziller, 1974.

Fuller, R. Buckminster. *Operating Manual for Spaceship Earth.* Southern Illinois Univ. Press, 1969.

Goodman. *People or Personnel: Decentralizing and the Mixed System.* New York: Random House, 1965.

Jones, Peter d'A. *The Consumer Society: A History of American Capitalism.* Baltimore: Penguin, 1965.

Kasson, John F. *Civilizing the Machine: Technology and Republican Values in America, 1776–1900.* New York: Penguin Books, 1977.

Kuhns, William. *The Post-Industrial Prophets.* New York: Harper & Row, 1971.

Madden, David, ed. *American Dreams, American Nightmares.* Carbondale: Southern Illinois Univ. Press, 1972.

Marx, Leo. *The Machine in the Garden: Technology and the Pastoral Ideal.* New York: Oxford Univ. Press, 1967.

Meier, Hugo A. "Technology and Democracy, 1800–1860." *Mississippi Historical Review,* 43 (March 1957).

Mumford, Lewis. *The Pentagon of Power.* New York: Harcourt, Brace, Jovanovich, Inc., 1970.

Ostrander, Gilman M. *American Civilization in the First Machine Age, 1890–1940.* New York: Harper Torchbooks, 1972.

Potter, David M. *People of Plenty: Economic Abundance and American Character.* Chicago: Univ. of Chicago Press, 1954.

"Technology in American Culture" [a special issue] *The American Examiner,* 5, No. 3 (Spring 1978).

West, Thomas Reed. *Flesh of Steel: Literature and the Machine in American Culture.* Charlotte, North Carolina: Vanderbilt Univ. Press, 1967.

II. *MACHINES, MYTHS, AND MARXISM*

Adorno, Theodor W. "Culture Industry Reconsidered." With an Introduction by Andreas Huyssen. *New German Critique,* 6 (Fall 1975), 3–19.

———. *Aesthetische Theorie.* Frankfurt am Main: Suhrkamp, 1970.

Benjamin, Walter. "The Artist as Producer." *Understanding Brecht.* London: NLB, 1973.

———. "The Work of Art in the Age of Mechanical Reproduction." *Illuminations.* New York: Harcourt, Brace & World, 1968.

Brüggemann, Heinz. *Literarische Technik und soziale Revolution.* Reinbek: Rowolt, 1973.

Bürger, Peter. *Theorie der Avantgarde.* Frankfurt am Main: Suhrkamp, 1974.

Enzensberger, Hans Magnus. *The Consciousness Industry.* New York: The Seabury Press, 1974.

Herff, Jeffrey. "Technology, Reification, and Romanticism." *New German Critique,* 12 (Fall 1977), 175–91.

Horkheimer, Max, and Theodor W. Adorno. *Dialectic of Enlightenment.* New York: The Seabury Press, 1975.

Jay, Martin. *The Dialectical Imagination*. Boston/Toronto: Little, Brown and Company, 1973.

Lethen, Helmut. *Neue Sachlichkeit 1924–1932*. Stuttgart: Metzler, 1970.

Negt Oskar. "Mass Media: Tools of Domination or Instruments of Liberation." *New German Critique*, 14 (Spring 1978), 61–80.

Real, Michael R. *Mass-Mediated Culture*. Englewood Cliffs, N.J.: Prentice-Hall, 1977.

Schiller, Herbert I. *The Mind Managers*. Boston: Beacon Press, 1973.

Winner, Langdon. *Autonomous Technology: Technics-Out-of-Control as a Theme in Political Thought*. Cambridge, Mass.: M.I.T. Press, 1977.

III. *SCIENCE FICTIONS*

Caronia, Antonio. "Incarnazioni dell'immaginario." In *Nei labirinti della fantascienza*. Milano: Feltrinelli, 1979, pp. 9–30.

Delany, Samuel R. *The American Shore: Meditations on a Tale of Science Fiction by Thomas M. Disch—Angouleme*. Elizabethtown, N.Y.: Dragon Press, 1978.

———. *The Jewel-Hinged Jaw*. Elizabethtown, N.Y.: Dragon Press, 1977; paperback ed. by Berkley-Windhover, 1978.

de Lauretis, Teresa. "SF in USA: linguaggio e corpo." *Alfabeta*, 3/4 (1979), 12–13.

Eizykman, Boris. "On Science Fiction." *Science Fiction Studies*, 2 (July 1975), 164–66.

Jameson, Fredric. "Generic Discontinuities in SF." *Science Fiction Studies*, 1 (Fall 1973), 57–68.

———. "World Reduction in LeGuin: The Emergence of Utopian Narrative." *Science Fiction Studies*, 2 (November 1975), 221–30.

Lem, Stanislaw. "On the Structural Analysis of Science Fiction." *Science Fiction Studies*, 1 (Fall 1973), 26–33.

Mullen, R. D., and D. Suvin, eds. *Science-Fiction Studies*. Boston: Gregg Press, 1976.

Nudelman, Rafail. "An Approach to the Structure of LeGuin's SF." *Science-Fiction Studies*, 2 (November 1975), 210–20.

Russ, Joanna. "Towards as Aesthetic of Science Fiction." *Science Fiction Studies*, 2 (July 1975), 112–19.

Suvin, Darko. "On the Poetics of the Science Fiction Genre." *College English*, 34 (1972), 372–82.

———. *Metamorphoses of Science Fiction: On the Poetics and History of a Literary Genre*. New Haven, Ct. and London: Yale Univ. Press, 1979.

Contributors

DAVID BATHRICK is Associate Professor of German at the University of Wisconsin-Madison and an editor of *New German Critique.* His publications include *The Dialectic and the Early Brecht* (1975) and numerous articles on twentieth-century literature, social theory, and East German culture and politics.

CHARLES CARAMELLO is Assistant Professor of English at the University of Maryland. He has co-edited *Performance in Postmodern Culture* (1977) with Michel Benamou and has published articles on modern and post-modern literature.

SAMUEL R. DELANY was Senior Fellow of the Center for Twentieth Century Studies in 1977. A science-fiction writer and critic, his works include the award-winning novels *Triton* (1976), *Dhalgren* (1977), *Nova* (1968) and *Babel—17* (1966) and two volumes of collected essays, *The Jewel-Hinged Jaw* (1978) and *The American Shore* (1979).

TERESA DE LAURETIS is Professor of Italian at the University of Wisconsin-Milwaukee, a member of the editorial board of *Yale Italian Studies* and an associate editor of *Ciné-Tracts.* Co-editor of *Theoretical Perspectives in Cinema* and *The Cinematic Apparatus,* she is the author of a book on the novelist Italo Svevo and of numerous essays in literary criticism, semiotics, and film theory published in Italy, Israel, Great Britain, Canada, and the United States.

HELEN FEHERVARY is Assistant Professor of German at Ohio State University and an editor of *New German Critique*. Her publications include articles on twentieth-century German literature, women's studies, and critical theory, and *Hölderlin and the Left: The Search for a Dialectic of Art and Life* (1977).

JOST HERMAND is Vilas Research Professor of German at the University of Wisconsin-Madison. His publications include the five volume study *Epochen deutscher Kultur von der Gründerzeit bis zur Gegenwart* (with Richard Hamann, 1959–1975), *Die Kultur der Weimarer Republik* (with Frank Trommler, 1978), and numerous other books and articles on nineteenth and twentieth-century German art, literature, and culture.

ANDREAS HUYSSEN is Associate Professor of German and Comparative Literature at the University of Wisconsin-Milwaukee and an editor of *New German Critique*. His publications include *Die frühromantische Konzeption von Uebersetzung und Aneignung* (1969), *Drama des Sturm und Drang* (1980), and numerous articles on twentieth-century literature and cultural theory.

JAMES MILLER is Assistant Professor in the Government Department of the University of Texas at Austin. He is the author of *History and Human Existence: From Marx to Merleau-Ponty* (1979) and the editor of *The Rolling Stone Illustrated History of Rock and Roll* (1976).

MILES ORVELL is Associate Professor of English at Temple University. In addition to articles on Nathanael West, Buster Keaton, and Flann O'Brien, he has written *Invisible Parade: The Fiction of Flannery O'Connor* (1972).

CARROLL PURSELL is Professor of History at the University of California, Santa Barbara. His many publications include *Readings in Technology and American Life* (1969), *The Military Industrial Complex* (1972), *From Conservation to Ecology: The Development of Environmental Concern* (1973), and, as co-editor, *Technology in Western Civilization*.

JAMES SCHMIDT is Assistant Professor of Government at the University of Texas at Austin and has published articles on critical theory, French Existentialism and modern social theory.

JOSEPH W. SLADE is Associate Professor of English and Director of the Communication Center at Long Island University-Brooklyn. He is the author of numerous articles on American literature and culture and of *Thomas Pynchon* (1974).

Darko Suvin is Professor of English at McGill University and is the author or editor of thirteen books on science fiction and literature. Co-editor of *Science Fiction Studies,* he is the vice-president of the Science Fiction Research Association.

Kathleen Woodward is Assistant Professor of English and Cultural and Technological Studies at the University of Wisconsin-Milwaukee. Her publications include *At Last, The Real Distinguished Thing: The Late Poems of Eliot, Pound, Stevens, and Williams* (forthcoming) and, as co-editor, *Aging and the Elderly: Humanistic Perspectives in Gerontology* (1978).